FOOD SCIENCE AND TECHNOLOGY

THERE'S THE BEEF

SELECT RESEARCH ON GLOBAL BEEF PRODUCTION AND TRADE

Food Science and Technology

Additional books in this series can be found on Nova's website
under the Series tab.

Additional E-books in this series can be found on Nova's website
under the E-book tab.

Agrilculture Issues and Policies

Additional books in this series can be found on Nova's website
under the Series tab.

Additional E-books in this series can be found on Nova's website
under the E-book tab.

FOOD SCIENCE AND TECHNOLOGY

THERE'S THE BEEF

SELECT RESEARCH ON GLOBAL BEEF PRODUCTION AND TRADE

JONATHAN L. RUIZ
AND
SEBASTIAN R. FOX
EDITORS

Nova Science Publishers, Inc.
New York

For permission to use material from this book please contact us:
Telephone 631-231-7269; Fax 631-231-8175
Web Site: http://www.novapublishers.com

Additional color graphics may be available in the e-book version of this book.

Library of Congress Cataloging-in-Publication Data

ISBN 978-1-61942-943-7

Published by Nova Science Publishers, Inc. † New York

CONTENTS

PREFACE

Beef cow-calf farms operate in an industry characterized by large numbers of small farms. Many of these farms specialize in beef cattle production, but farm households on these operations tend to generate more income from off-farm sources, such as wages and salaries or retirement income, than from the farm businesses themselves. This book discusses select research on global beef production and trade including beef production in the United States, Japan and Mexico.

Chapter 1 - Beef cow-calf production in the United States is widespread, occurring in every State. Nearly 765,000 farms, about 35 percent of the 2.2 million farms in the United States, had a beef cow inventory in 2007. Most of these were small, part-time operations. About a third of farms that raise beef animals had a beef cow inventory of less than 10 cows, more than half had fewer than 20 cows, and nearly 80 percent had fewer than 50 cows. In this study, ERS uses data from USDA's 2008 Agricultural Resource Management Survey for U.S. beef cow-calf operations to examine the structure, costs, and characteristics of beef cow-calf producers. Many small operations are "rural residence farms" that specialize in beef cow-calf production, but their income from off-farm sources exceeds that from the farm. Most beef cow-calf production occurs on large farms, but cow-calf production is not the primary enterprise on many of these farms. Findings suggest that operators of beef cow-calf farms have a diverse set of goals for the cattle enterprise.

Chapter 2 - Fed cattle from the Southern Plains generally are priced at a premium to Northern and Central Plains cattle. The "North-South price spread" persists because, while production practices in the US cattle sector typically include pasture grazing followed by feeding grain in feedlots, the timing and the duration between each stage may vary. Regional differences include variation with the length and timing of the calving season, the availability and quality of pasture (including the potential for winter wheat grazing), and grain prices. This report describes nuances in regional production practices that underlie the price relationship referred to as "the North-South spread."

Chapter 3 - Animal byproducts contribute to the bottom line of the U.S. meat industry. Byproducts (edible offal (including variety meats), inedible offal, hides and skins, blood, fats, and tallow) include all parts of a live animal that are not part of the dressed carcass and constitute about 30 percent of the liveweight of hogs and about 44 percent of the live-weight of cattle.

Byproducts from animal slaughter provide raw materials used in pharmaceutical, cosmetic, household, and industrial products.

Exports of edible offal also contribute to the value and profitability of the U.S. meat processing industry in a way that leads to higher prices for livestock producers, as byproducts account for more than 23 and 35 percent of the volume of beef/veal and pork exports, respectively.

Regression analysis indicates that a 10-percent increase in the steer byproduct drop value adds a 1-percent increase in the five-area weighted average price for all grades of steers.

U.S. exports of beef/veal and pork edible offal have increased in recent years, mostly due to population growth, income growth, and consumer preferences for variety meats, especially in Asia. Income growth in the global marketplace, however, may have varied effects on the consumption and trade of variety meats.

Chapter 4 - Japan imports large amounts of beef, primarily from Oceania and North America, and its consumers are willing to pay a premium for heavily marbled, grain-fed beef. As a result, Japan's import of certain beef cuts and offal raise the value of U.S. cattle. Trade bans resulting from the discovery of bovine spongiform encephalopathy (BSE) in North America dramatically shifted beef supplies to imported beef from Australia and New Zealand.

Beef consumption in Japan may increase from current levels in Japan's market, particularly if prices fall or income rises, despite a declining population. Economic factors, demographic factors, import and domestic policies and regulations, as well as consumer tastes and preferences, will determine the outlook for beef consumption in Japan and the ability of U.S. beef to compete in that market.

Japan's domestic beef production relies upon imported feed, primarily from the United States, to feed specific breeds energy-intensive rations. Government support plays an important role in the market structure of the domestic beef industry, which has an impact on imported beef from other countries, including the United States.

Chapter 5 - Cattle raised for export in Mexico represent, on average, over half of total U.S. cattle imports, as the United States has the comparative advantage in feeding cattle. In turn, Mexico imports the largest quantities of U.S. beef and beef products of all U.S. beef trading partners to satisfy the increasing demand for beef, including higher quality feedlot-finished beef.

The authors describe the production systems and stocker and feedlot production practices commonly employed in Mexico, including regional factors like production parameters, cattle marketing, and sourcing practices. Changes in consumer demand and continued population and economic growth support the potential for increased fed beef demand in Mexico.

As a result of increased fed beef production, feed demand from other livestock industries, competition from crop production for land and water to produce feed and forage, and international competition for inputs, the Mexican beef industry will face tremendous challenges in the coming decades.

Chapter 6 - Mexico often exports over a million head of cattle annually to the United States. These animals, primarily intended for U.S. stocker and/or feeder operations, all pass through a cow-calf production system of one form or another in Mexico. This report outlines cow-calf production types, beef production systems, and geo-climatic influences on beef production in Mexico. The authors of this report also evaluate factors affecting Mexican beef production, such as patterns of landownership, major diseases and parasites in cattle production, changes in export regions, and changes in domestic consumption as they pertain to cow-calf production. Amid changes in consumer demand and the continuous challenges pertaining to Mexico's cattle export market, production practices at the cow-calf level may pinpoint opportunities and constraints within the domestic beef markets of both Mexico and the United States.

In: There's The Beef
Editors: Jonathan L. Ruiz and S. R. Fox
ISBN 978-1-61942-943-7

Chapter 1

THE DIVERSE STRUCTURE AND ORGANIZATION OF U.S. BEEF COW-CALF FARMS*

William D. McBride and Kenneth Mathews, Jr.

ABSTRACT

Beef cow-calf production in the United States is widespread, occurring in every State. Nearly 765,000 farms, about 35 percent of the 2.2 million farms in the United States, had a beef cow inventory in 2007. Most of these were small, part-time operations. About a third of farms that raise beef animals had a beef cow inventory of less than 10 cows, more than half had fewer than 20 cows, and nearly 80 percent had fewer than 50 cows. In this study, ERS uses data from USDA's 2008 Agricultural Resource Management Survey for U.S. beef cow-calf operations to examine the structure, costs, and characteristics of beef cow-calf producers. Many small operations are "rural residence farms" that specialize in beef cow-calf production, but their income from off-farm sources exceeds that from the farm. Most beef cow-calf production occurs on large farms, but cow-calf production is not the primary enterprise on many of these farms. Findings suggest that operators of beef cow-calf farms have a diverse set of goals for the cattle enterprise.

Keywords: Beef cow-calf production, farm income, animal traceability, Agricultural Resource Management Survey (ARMS), National Animal Identification System (NAIS)

SUMMARY

What Is the Issue?

Beef cow-calf farms operate in an industry characterized by large numbers of small farms. Many of these farms specialize in beef cattle production, but farm households on these

---

* This is an edited, reformatted and augmented version of TheUnited States Department of Agriculture publication, Economic Information Bulletin Number 73, dated March 2011.

operations tend to generate more income from off-farm sources, such as wages and salaries or retirement income, than from the farm businesses themselves. Large farms account for most beef cow-calf production in the United States, but on many of these farms, cow-calf production is not the primary enterprise. These findings suggest that operators of beef cow-calf farms, large and small, have varying goals for their cattle enterprises, of which farming as a lifestyle choice is not uncommon.

What Did the Study Find?

- About 60 percent of U.S. beef cow-calf farms produce calves that are sold at or shortly after weaning. These are usually small farms, and most are located in the Southeast and Southern Plains. Many of the farm households on these operations generate most of their income from off-farm sources.
- More than a third of beef cow-calf farms retain ownership of calves after weaning and continue grazing, or backgrounding, the calves from 30 to 90 days before selling. These farms are generally larger, have more beef cows, and are distributed throughout the United States, with many in the Northern Plains and West regions.
- The majority of U.S. beef cows are located in the South, including the Southern Plains (primarily Texas) and the Southeast. These regions have the advantage of a longer grazing season and less need for supplemental forage to support beef cattle during the winter, which results in lower feed costs. Despite higher feed costs in the Northern Plains, large beef cow-calf producers in this region are able to compete with those in the South due to production efficiencies and economies of size.
- Economies of size in beef cow-calf production suggest that farms have an incentive to become larger. However, the significant land area required for large-scale beef cow-calf production inhibits many producers from expanding. In most areas of the United States, beef cow-calf production is the residual user of land. As the opportunity cost of pasture and range land increases for uses such as crop production or recreational activities, the size of beef cow-calf operations may be limited or fragmented into smaller units.
- Most farms with beef cows do not specialize in beef cow-calf production. In 2008, cattle production accounted for less than 40 percent of the average farm product value on U.S. beef cow-calf farms. Regionally, cattle production accounted for about two-thirds of farm product value on beef cow-calf farms in the Southern Plains and West regions but less than 40 percent in other regions. Specialization in cattle production increased with farm size and peaked at 60 percent of farm product value for operations with 250-499 beef cows. Among the largest operations, those with 500 or more cows, less than 50 percent of farm product value was from cattle.
- Operators of more than a third of beef cow-calf farms worked off-farm in 2008, and half of beef cow-calf farms are classified as rural residence farms. These farms are small operations that specialize in beef cow-calf production but report off-farm earnings as the primary source of household income. Commercial farms with beef cow-calf enterprises are mostly diversified farm operations on which cattle are a secondary enterprise that accounts for about a fourth of farm product value. On inter-

mediate farms, which have annual farm sales under $250,000 and report farming as the main occupation, the beef cattle enterprise accounts for over half of farm product value. Intermediate farms are among the most financially vulnerable to the input and output price variations of beef cattle production.

- In 2008, more than 80 percent of beef cow-calf producers had some type of animal identification system in place, such as branding or ear tagging. But, nearly a quarter of beef cow-calf producers with 20 or more cows reported a lack of familiarity with the National Animal Identification System (NAIS), and only about a quarter had their premises registered with the system. This lack of participation among the Nation's nearly 765,000 beef cow-calf producers appears to be related to concerns about liability and costs associated with the program. Because beef cow-calf production is a secondary farm enterprise and a secondary household income source for most farms with beef cows, there may be little incentive for these farms to risk any perceived liability or to incur program participation costs. This may create a challenge for Federal or State efforts to enhance product traceability through animal identification on beef-cow calf farms.

How Was the Study Conducted?

In this report, ERS summarizes information from a 2008 survey of U.S. beef cow-calf producers included as part of USDA's annual Agricultural Resource Management Survey, which is administered by ERS and USDA's National Agricultural Statistics Service. The survey covered 22 States and targeted beef cow-calf producers with at least 20 beef cows on the operation during 2008. Data from participating producers were weighted for analysis such that they represent 96 percent of U.S beef cow-calf farms in the target population. Surveyed producers were divided into groups by type of operation (cow-calf only, cow-calf/stocker, or cow-calf/feedlot), region, size, and farm typology, through which structural and economic differences among producers in each group were statistically evaluated. Beef cow-calf producers were also grouped according to their familiarity and participation with the NAIS. Program participants were statistically contrasted with other producers to identify characteristics that distinguish those who participated in animal identification and product traceability programs.

INTRODUCTION

Beef cow-calf operations are located throughout the United States, typically on land not suited or needed for crop production. These operations depend on range and pasture forage conditions, which are in turn dependent upon variations in the average level of rainfall and temperature for an area. Beef cows harvest forage from grass and range lands to maintain themselves and raise a calf with little grain fed. Forage availability determines the stocking capacity of range and pasture land and can determine whether calves are sold at weaning or retained for additional grazing and growth (Hoder et al., 2007).

Beef cow-calf production contributes to the agricultural economy in most of the United States. According to the Census of Agriculture, cattle and calf sales generated about $61.2 billion in 2007,[1] accounting for about 20 percent of the total market value of agricultural products sold in the United States during the period and ranking first in sales among all commodities. Nearly 765,000 farms, about 35 percent of the 2.2 million farms in the United States, had a beef cow inventory in 2007 (table 1). Most of these were small operations. About a third of farms had a beef cow inventory of less than 10 cows, more than half had fewer than 20 cows, and nearly 80 percent had fewer than 50 cows (USDA, NASS, 2007).

Rapid transformation in the size, technologies used, and ownership structure has characterized operations in most livestock sectors during the past decade. For example, the number of hog and dairy farms in the United States fell by around 40 percent between 1997 and 2007 (table 1), while the average size of these operations nearly doubled. In contrast, the number of operations with beef cows fell only 15 percent between 1997 and 2007, and the average size of these farms increased from 38 to 43 head, or about 13 percent. Beef cow-calf operations are primarily tied to land suitable for grazing cattle, while hog and dairy operations have substituted capital for land, facilitating large-scale production by moving animals into confinement facilities.

Feeder cattle prices and supplemental feed costs are important factors affecting the profitability of beef cow-calf production. Feeder cattle prices are affected by prices paid for slaughter cattle, which in turn, are affected by consumer demand for beef as reflected in retail beef prices. Cow-calf producers respond to high (low) feeder cattle prices by increasing (decreasing) production, but biological constraints of cattle prevent producers from instantly responding to price. This gives rise to the cattle cycle— cyclical increases and decreases in the cattle herd over time, determined by the combined effects of cattle prices; the time needed to breed, calve, and raise cattle to market weight; and climatic conditions (USDA, ERS, 2010b).

Beef cow-calf production costs are sensitive to the amount of supplemental feed, in addition to grazing, that is required to over-winter, feed during drought, or otherwise maintain beef cows during the year. There is a cost advantage for producers in areas with a longer grazing season and milder winter, thus requiring less supplemental feed. Other factors shown to influence beef cow-calf production costs are investments in machinery and equipment, calving percentage, calf death loss, and length of the breeding season (Ramsey et al., 2005).

This report presents information about U.S. beef cow-calf operations, emphasizing the diverse structural and economic characteristics of the farms and ranches. The objective is to provide updated information about the type of farms, size of farms, farm production practices, and farm operator and financial characteristics of the U.S. beef cow-calf industry.

Table 1. Number of U.S. farms and farms with livestock operations

Type of farms	Number of farms		Percent change
	1997	2007	
All farms	2,215,876	2,204,792	-0.5
Farms with beef cows	899,756	764,984	-15.0
Farms with milk cows	125,041	69,890	-44.1
Farms with hogs and pigs	124,889	75,442	-39.6

Source: USDA, Economic Research Service using 2007 Census of Agriculture.

USDA's Agricultural Resource Management Survey

This report summarizes data from an indepth survey of U.S. beef cow-calf producers conducted as part of USDA's 2008 Agricultural Resource Management Survey (ARMS). ARMS is administered annually by ERS and the National Agricultural Statistics Service (NASS). Survey responses were collected from a cross-section of U.S. beef cow-calf operations and include measures of farm size, production costs, business arrangements, production facilities and practices, and farm operator and financial characteristics. The sampling resulted in 1,966 usable responses from 3,600 surveyed producers in 22 States (fig. 1), a survey response rate of 55 percent. Producers in the Northeast, Upper Midwest, and other States were not surveyed because of their minor share of U.S. beef cow-calf production and because of survey cost limitations.

Beef cow-calf farms surveyed in the 2008 ARMS were chosen from a list of farm operations maintained by USDA's NASS. The survey's target population was farms with 20 or more beef cows on the operation at any time during 2008. A primary purpose of the beef cow-calf survey was to collect farm characteristics, farm financial data, and commodity costs-of-production for commercial beef cow-calf operations. Farms with fewer than 20 cows were screened out to exclude farms that raise cattle for onfarm consumption and other noncommercial activities, such as youth projects or hobby farming. Most U.S. beef cow-calf farms are small operations, and limiting the target population to those with fewer than 20 cows omits about 53 percent of the farms with beef cows reported in the 2007 Census of Agriculture. These smallest farms, however, accounted for only about 10 percent of the U.S. beef cow inventory in 2007.

Each farm surveyed in ARMS represents a number of similar farms in the population as indicated by the survey expansion factor.

The expansion factor, or survey weight, was determined from the farm's selection probability and thereby expands the sample to represent the target population.[2]

Compared with data in the 2007 Census of Agriculture, the sample represents approximately 96 percent of U.S. beef cow-calf farms in the ARMS target population (20 or more cows) and 93 percent of the beef cow inventory on these farms.[3]

The difference is primarily due to ARMS collecting information from operators in only 22 States. Figure 2 displays data on beef cow-calf farms and beef cow inventories from the 2007 Census of Agriculture and the 2008 ARMS by farm size category. Both beef cow-calf farms and beef cow inventories in the ARMS sample are distributed across the size categories much like those in the Census of Agriculture.

Other Data

ERS analysts relied on two other sources of data on farms with beef cows. Data from the 1997, 2002, and 2007 Censuses of Agriculture were used to measure changes in farm numbers and cow inventories that provide a context for the ARMS data. These data provide the most reliable estimates of the number of farms and beef cows.

Data from USDA's National Animal Health Monitoring System (NAHMS) were also used to provide context for the ARMS data. NAHMS data provide indications of production practices used on beef cow-calf farms. The NAHMS study of beef cow-calf producers was conducted for 2007 and 2008 and provides information about the health, production, and management practices of beef cow-calf operations. NAHMS coverage closely aligns with, and can be considered complementary to, ARMS. The inference population for the NAHMS study is operations with 1 or more beef cows located in the same 22 States surveyed in ARMS, plus Idaho.

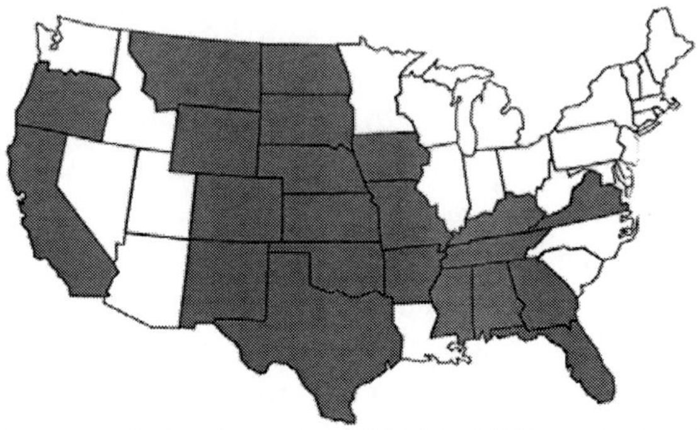

Source: USDA, Economic Research Service using USDA's 2008 Agricultural Resource Management Survey (ARMS).

Producers in the surveyed States (shaded) accounted for about 96 percent of beef cow farms with 20 or more head and 93 percent of the beef cow inventory on these farms.

Figure 1. States surveyed in the 2008 ARMS of U.S. beef cow-calf producers.

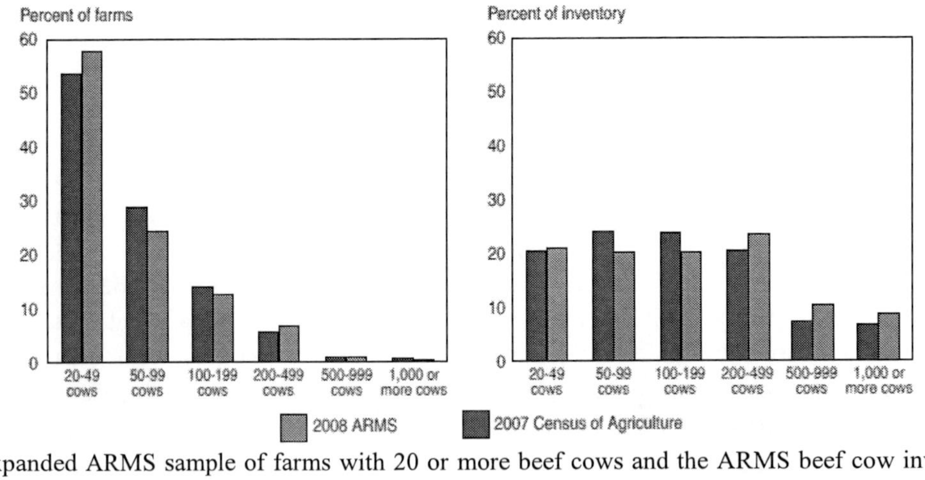

The expanded ARMS sample of farms with 20 or more beef cows and the ARMS beef cow inventory are distributed similarly across identical size categories of the 2007 Census of Agriculture.

Source: USDA, Economic Research Service using 2007 Census of Agriculture; and USDA's 2008 Agricultural Resource Management Survey (ARMS).

Notes: The 2007 Census of Agriculture measured beef cow inventory on December 31, 2007. The ARMS beef cow inventory is that on January 1, 2008. Differences between the estimates are primarily due to fewer States included in ARMS, and the sampling and nonsampling error of ARMS. Limiting ARMS to operations with 20 or more beef cows omitted about 53 percent of farms with beef cows and 10 percent of the beef cow inventory reported in the 2007 Census of Agriculture.

Figure 2. Survey coverage of U.S. beef cow-calf farms and inventory by size of operation, 2008.

Classifying Beef Cow-Calf Producers

Commercial beef cattle production in the United States can be classified into three phases: (1) cow-calf—cow maintenance during breeding, gestation, and calving to when calves are weaned from between 6 and 9 months of age, weighing 400-700 pounds; (2) stocker—addition of 200-400 pounds of extra weight to weaned calves over 3-8 months; and (3) feedlot—finishing of calves, usually on a combination of forage and grain, to a slaughter weight of 1,000-1,500 pounds (see box, "A Primer on Beef Cow-Calf Production"). About half of U.S. beef cattle operations specialize in the cow-calf phase (USDA, APHIS, 2010a), but the remainder conduct activities in two or all three of the phases. It is common for beef cow-calf producers to retain calves after weaning and conduct the stocker phase on the same operation, often referred to as back-grounding calves (see glossary).

Farm and Operator Characteristics by Producer Classification

The average cow-calf operation surveyed in the 2008 ARMS had a peak inventory of 102 beef cows, had an average inventory of 79 cows, and weaned 73 calves during 2008 (table 2).

A Primer on Beef Cow-Calf Production

In most of the United States, beef cows are bred during the summer in order to calve 9 months later in late winter or early spring. This system takes advantage of abundant summer pasture for cattle grazing at a time when lactating cows have their greatest nutritional requirements. In midwinter, most cows are not lactating and thus have lower nutritional requirements. Late winter calving also fits into a slack labor period on most farms.

Fall calving is used on some operations. In this system, calves are born in mid to late fall and marketed anywhere from late spring to early summer. An advantage of this system is that calves are old and large enough by spring and early summer to utilize grass pastures. Fall calves are typically heavier at weaning than spring calves, but the greater cost of feeding a lactating cow through the winter may offset any additional value from the heavier calf. Fall calving may also interfere with harvesting field crops on some farms.

The ideal time of year for calving on a beef cow-calf operation depends on the forage and/or feed supply, available labor, and the intended marketing dates. Also important is a controlled, scheduled calving season (60-90 days), as opposed to year-round calving. With a controlled, scheduled calving season, such as spring or fall calving, (1) most herd management practices can be performed at the same time, (2) use of time and labor can be concentrated and efficient, (3) nonbreeding cows can be more easily identified, and (4) a more uniform calf crop can be produced. Despite these advantages, National Animal Health Monitoring System (NAHMS) data indicate that a significant number of U.S. operations do not use seasonal calving. About 46 percent of U.S. beef cow-calf operations reported year-round calving (USDA, APHIS, 2010b).

Calves are weaned at 400-700 pounds, at 6-9 months of age. Calves are either sold directly after weaning or kept on the operation in a preconditioning program, often referred to as backgrounding, before being sold. Keeping calves on the operation after weaning eliminates the stress of transportation and provides an opportunity to acclimate calves to eating from a feed bunk. NAHMS data show that about half of the calves in the United States are sold at weaning, and more than 80 percent are sold within 60 days of weaning (USDA, APHIS, 2010a).

Table 2. Farm and operator characteristics of U.S. beef cow-calf farms, by producer classification, 2008

Item	Cow-calf only(a)	Cow-calf/ stocker(b)	Cow-calf/ feedlot (c)	All beef cow -calf farms
Percent of farms/beef cows	47/36	44/53	9/10	100/100
Beef cows—peak per farm[1]	79^{bc}	123^{a}	116^{a}	102
Beef cows—average per farm[2]	64^{bc}	93^{a}	86^{a}	79
Beef cows calving	69^{bc}	106^{a}	97^{a}	88
Calves weaned	57^{bc}	88^{a}	82^{a}	73
Weaning age (days)	210	209	210	210
Weaning weight (lbs)	502^{c}	499^{c}	523^{ab}	502
Percent of calves:				
Sold at weaning	100^{bc}	21^{ac}	28^{ab}	59
Backgrounded then sold	0	79^{c}	15^{b}	36
Retained until slaughter	0	0	57	5
Total farm production value ($)	$93,568^{bc}$	$144,305^{ac}$	$279,905^{ab}$	132,794
Percent from cattle	36^{b}	43^{ac}	34^{b}	39
Acres operated	$1,007^{bc}$	$1,623^{a}$	$1,407^{a}$	1,316
Private pasture/range acres	792^{b}	$1,274^{ac}$	969^{b}	1,019
Acres per beef cow	10	11c	8^{b}	10
Percent using private pasture/range	94^{b}	91^{a}	91	93
Percent using public grazing land	3bc	6a	8a	5
Crops produced (percent of farms):				
Corn	6^{bc}	20^{ac}	44^{ab}	16
Soybeans	5^{bc}	14^{ac}	43^{ab}	13
Small grains	10^{bc}	17^{a}	21^{a}	14
Hay	77	78	79	78
Location (percent of farms):				
North Central	13^{bc}	16^{ac}	28^{ab}	16
Southeast	39^{bc}	28^{ac}	12^{ab}	32
Northern Plains	9^{bc}	21^{ac}	31^{ab}	16
Southern Plains	33^{bc}	24^{ac}	14^{ab}	27
West	6^{bc}	11^{a}	14^{a}	9
Operator:				
Age (years)	61^{bc}	59^{ac}	56^{ab}	60
Age (percent greater than age 65)	42^{bc}	34^{ac}	21^{ab}	36
Off-farm occupation (percent)	41^{bc}	34^{ac}	20^{ab}	36
Completed college (percent)	23^{b}	29^{a}	27	26
Exit within 5 years (percent)	26^{bc}	22^{a}	18^{a}	23

Source: USDA, Economic Research Service using data from USDA's 2008 Agricultural Resource Management Survey (ARMS).

Notes: The superscripts refer to the results of statistical tests of differences between item means in each column. All tests are expressed at a 95-percent level of confidence. A lettered superscript denotes that the item mean reported in a column is significantly different from that in the superscript column. The tests were conducted using a jackknife variance estimator with 15 replicates provided with the ARMS data.

[1] Largest number of beef cows and heifers that have calved on farm at any time during 2008.

[2] Average of beginning and ending inventories of beef cows and heifers that have calved on farm during 2008.

Some cow-calf producers also purchase calves for backgrounding, and weaned calves may be backgrounded on specialized stocker operations. The decision to sell calves at weaning or to retain ownership or purchase calves for backgrounding is made each year based on cattle prices and forage availability. Cow-calf/stocker producers, acting as speculators, distribute the seasonal inventory of calves on the market across the year.

Relatively few cow-calf operations finish calves in farm feedlots. Instead, most calves are finished in large commercial feedlots that purchase most or all feed ingredients, employ nutritionists and veterinarians, and buy and sell cattle weekly (MacDonald and McBride, 2009). However, beef cow-calf producers can also retain ownership of the calves being finished in commercial feedlots.

NAHMS data about days that calves were held on the operation after weaning provide an indication of the production phases conducted on farms with beef cows. Half of all beef cow-calf operations in NAHMS held calves zero days after weaning (USDA, APHIS, 2010a). Selling calves at weaning was most common among small operations and those in the South Central and Southeast regions of the United States. Ten percent of operations retained calves for 123 days or more. This characteristic was more common among large operations and those in the West. Beef cow-calf producers in ARMS were classified according to which phases were conducted on the farm. *Cow-calf only* operations reported that all calves were sold at weaning. This producer classification accounted for 47 percent of farms in ARMS and 36 percent of beef cows (table 2). *Cow-calf/stocker* operations reported that they retained calves after weaning but sold the calves before finishing. This classification accounted for 44 percent of surveyed farms and 53 percent of beef cows.

Cow-calf/feedlot operations, including those reporting that they retained ownership of the calves until slaughter, accounted for only 9 percent of farms and 10 percent of beef cows. All calves on the cow-calf only operations were sold at weaning, while nearly 80 percent of the calves weaned on cow-calf/stocker operations were retained after weaning, backgrounded, and sold as stockers.

Cow-calf/feedlot farms were concentrated in the North Central and Northern Plains regions, together accounting for nearly 60 percent. Cow-calf/feedlot farms were most often located in areas where corn and other crops are readily available for finishing cattle, while cow-calf only farms were located more often in areas with abundant pasture or range land.

Beef cow-calf producers averaged 60 years of age, and more than a third had a primary occupation off-farm (see table 2). Operators of cow-calf only farms were older than operators of other farm types and more often worked off-farm. More than 40 percent of cow-calf only farm operators were olderthan age 65, compared with 34 and 21 percent on cow-calf stocker and cowcalf/feedlot farms, respectively. More than 40 percent of cow-calf only farm operators had a primary occupation off-farm, compared with 34 percent on cow-calf/stocker farms and 20 percent on cow-calf/feedlot farms. With fewer beef cows on cow-calf only farms, labor requirements were reduced and farm operators were available to work more time off-farm.

Production Practices and Farm Finances by Producer Classification

The technologies and practices used on beef cow-calf farms were found to vary substantially. Cow-calf only producers were less likely than other cow-calf producers to use many beef cow-calf production practices (table 3), including a defined calving season, artificial insemination, growth promoting implants and/or ionophores, and veterinary and nutritional services. Information management technologies (see glossary), including onfarm computer recordkeeping systems and the Internet, were also used less often on cow-calf only operations than on cow-calf/stocker and cow-calf/feedlot operations.

Differences in the use of management technologies may be due to time available for cow-calf production given that more operators of cow-calf only farms worked primarily off-farm (Fernandez-Cornejo, 2007).

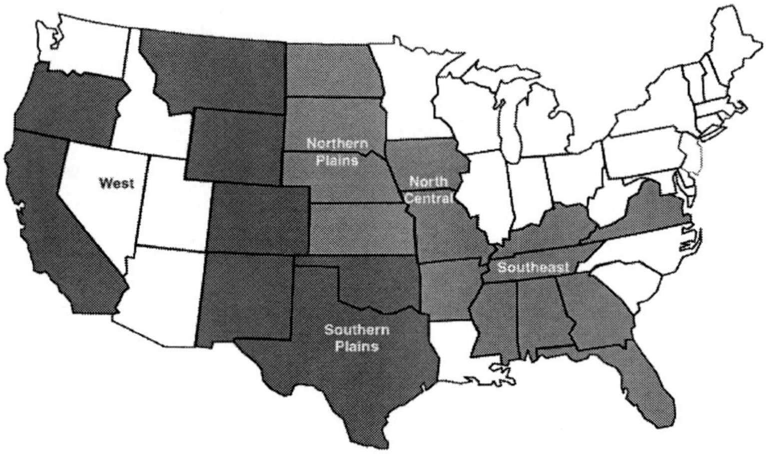

Source: USDA, Economic Research Service using USDA's 2008 Agricultural Resource Management Survey (ARMS).

Nearly a third of ARMS cow-calf producers were located in Southeast States, and 27 percent were in the Southern Plains region.

Figure 3. U.S. beef cow-calf production regions.

U.S. beef cow-calf operations were generally part of diversified farm operations in 2008 as average gross cash income per farm was nearly $48,000 from crop sales and about $58,000 from cattle and other livestock sales (table 3). Among the producer types, cow-calf/feedlot farms were the largest and most diversified, with annual gross cash income more than $100,000 higher than on other farms.[6]

Cow-calf feedlot producers benefitted from high prices for corn and soybeans in 2008 as crop sales were greatest on these farms, accounting for 46 percent of cash income. On cow-calf/stocker farms, cattle sales accounted for more than 40 percent of gross cash income. Both crop and cattle sales were lower (and roughly the same) on cow-calf only farms than on other farm types. Cow-calf/feedlot farms had the highest average net cash ($48,000) and farm income ($32,000), about triple that on cow-calf only farms. Net cash and farm income were lowest on cow-calf only farms, both averaging less than $15,000 for 2008.

The amount of off-farm income earned by beef cow-calf producers in 2008 dwarfed farm income. Average off-farm income was about $72,000 among all beef cow-calf farms, nearly three times higher than net cash farm income and $20,000 more than gross cattle sales.

Off-farm income was more than twice gross cattle sales on cow-calf only farms and compensated for the low level of farm income. Off-farm income was a substantial part of household income for each beef cow-calf producer classification.

Regional Diversity in Beef Cow-Calf Production

Although beef cows are kept on range or pasture land throughout the year in much of the Nation, few producers can depend on grazing alone to furnish a year-round supply of forage. In parts of the country, snow cover occasionally prevents grazing during the winter. In other parts, growth and nutritive content of range and pasture plants limit grazing during part of the year. Thus, most producers—97 percent according to NAHMS—feed cows some harvested forage almost every year (USDA, APHIS, 2010b). These harvested forages are a substantial production cost for producers in some areas of the United States (USDA, ERS, 2010d).

Table 3. Production practices and financial characteristics of U.S. beef cow-calf farms, by producer classification, 2008

Item	Cow-calf only(a)	Cow-calf/stocker(b)	Cow-calf/feedlot(c)	All beef cow-calf farms
Practices:		*Percent using*		
Defined calving season	54[bc]	66[ac]	79[ab]	61
Commercial seed stock	16[b]	13[ac]	20[b]	15
Artificial insemination	4[bc]	11[ac]	19[ab]	8
Calf implants and/or ionophores	9[bc]	17[ac]	25[ab]	14
Regular veterinary services	17[bc]	26[a]	32[a]	23
Nutritionist services	4bc	8ac	18[ab]	7
Rotational grazing	59	62	56	60
Tested forage quality	12[bc]	19[ac]	27[ab]	16
Individual cow records	40[bc]	50[a]	56[a]	46
Onfarm computer records	17[bc]	22[ac]	29[ab]	20
Internet for cattle information	29[bc]	38[a]	42[a]	34
Forward purchase inputs	5[bc]	10[a]	12[a]	8
Negotiate input prices	14[bc]	23[a]	24[a]	19
Farm finances:		*Dollars per farm*		
Gross cash income[1]	74,894[bc]	143,793[ac]	254,531[ab]	121,574
-Cattle sales	33,613[bc]	62,384[ac]	94,180[ab]	51,809
-Other livestock sales	1,269[bc]	10,543[a]	14,000[a]	6,525
-Crop sales	30,620[bc]	51,451[ac]	117,514[ab]	47,657
Cash expenses	59,918[bc]	113,200[ac]	206,892[ab]	96,740
-Variable	46,221[bc]	86,250[ac]	158,893[ab]	74,077
-Fixed	13,698[bc]	26,950[ac]	48,099[ab]	22,663
Net cash farm income	14,975[bc]	30,594[a]	47,640[a]	24,834
Net farm income[2]	11,972[bc]	24,292[a]	31,595[a]	19,197
Off-farm income[3]	69,019	77,465[c]	59,975[b]	71,947

Source: USDA, Economic Research Service using data from USDA's 2008 Agricultural Resource Management Survey (ARMS).

Notes: The superscripts refer to the results of statistical tests of differences between item means in each column. All tests are expressed at a 95-percent level of confidence. A lettered superscript denotes that the item mean reported in a column is significantly different from that in the superscript column. The tests were conducted using a jackknife variance estimator with 15 replicates provided with the ARMS data.

[1] In addition to livestock and crops sales, gross cash income includes Government payments and other farm-related income.

[2] Net farm income equals net cash income less depreciation and the value of noncash benefits for hired labor, plus the net value of commodity inventory change and nonmonetary income.

[3] Off-farm income includes farm household earnings from wages and salaries, earnings from another business, interest and dividend income, pensions and other retirement income, and income from other off-farm sources.

Cow-calf herds are grazed not only on range and pasture lands but also on land primarily used for other purposes. In some areas, wheat can be grazed for a period during the fall and winter before being harvested for grain later in the year. Crop residues are also grazed in some areas for limited periods following the harvest of corn, grain sorghum, and other crops. Feed costs, including purchased feed, harvested forages, and grazing, frequently account for about two-thirds of the total operating costs of beef cow-calf production (USDA, ERS, 2010d).

Regional variation in climate and grazing conditions affects the calving and weaning practices of beef cow-calf producers. NAHMS data show that a majority of beef cow-calf farms in Southern States calved year-round.[7] In Western States, nearly 80 percent used a spring calving system (USDA, APHIS, 2010b), as calving was scheduled to avoid severe winter weather in much of the West and to take advantage of forage conditions in the summer and on public grazing land (see glossary). More than 60 percent of cow-calf farms in the South sell calves at weaning, compared with 36 percent in the West and 26 percent in the central region (USDA, APHIS, 2010a).

Regional diversity in beef cow-calf production is examined in this report by summarizing farm and operator characteristics and farm production practices and finances for producers in the five beef cow-calf production regions shown in figure 3. Among the regions, little has changed in the inventory of beef cows since 1997 (fig. 4).

Regions in the South—the Southern Plains and Southeast—have the most beef cows, and Texas, with more than 5 million head, has more than twice the number of beef cows as any other State. The North Central region has the fewest beef cows, but Missouri ranks second among States in the number of beef cows with more than 2 million head.

Farm and Operator Characteristics by Region

Nearly 60 percent of beef cow-calf farms surveyed in ARMS were in the Southeast and Southern Plains regions, along with 50 percent of beef cows (table 4). Beef cow-calf farms in the Northern Plains and West together had nearly 40 percent of beef cows, despite accounting for only 25 percent of farms. The largest beef cow-calf operations were in the Northern Plains and West, averaging 129 and 213 cows per farm (at peak), respectively, and the smallest were in the North Central and Southeast, averaging 65 and 78 cows (at peak), respectively. About 70 percent of the calves from beef cow-calf operations in the Southeast and Southern Plains were sold at weaning, compared with about 50 percent or less in the other regions where weaned calves were more often backgrounded before being sold. This may explain the older weaning age and greater weaning weight of calves in the Northern Plains and West regions.

Farm value of production was highest for cow-calf farms in the Northern Plains, at an average of more than $200,000 per farm, followed by that for farms in the West (table 4). Cattle production accounted for a significantly larger part of farm product value (about two-thirds) in the West and Southern Plains than in other regions. Acres operated were also much larger in the West, averaging over 4,000 per farm, more than twice that in any other region.

Nearly all of this acreage was private pasture and range land for grazing cattle. In addition to using private pasture and range land, more than a third of farms in the West used public grazing land. Cattle production accounted for 25 percent or less of farm product value on beef cow-calf farms in the North Central and Southeast.

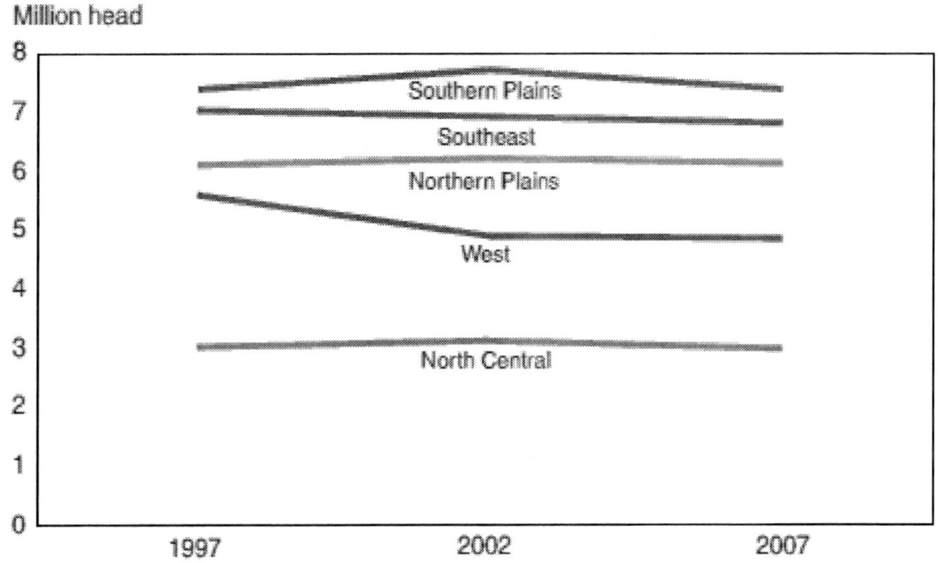

Source: USDA, Economic Research Service using 2007 Census of Agriculture.
Regions in the South, Southern Plains, and Southeast had the most beef cows each year, and the number of beef cows in each region has changed little over time.

Figure 4. Beef cow inventory by U.S. production region.

More than 90 percent of farms in most regions used private pasture land for grazing beef cattle, although the acreage and stocking rate varied significantly among the regions. Farms in the Northern Plains and Southern Plains averaged more than 2,000 and 1,400 acres in size, respectively, with much of this in pasture and range land for cattle. Farms in the North Central and Southeast had the fewest acres, 518 and 453, respectively, each having just over 200 acres of pasture. Still, the pasture acreage in these regions supported more beef cows (about 3 acres per cow) than that in other regions (10-20 acres per cow).

More than 80 percent of farms in the North Central, Southeast, and Northern Plains produced hay for cattle feed, compared with about 60 percent in the other regions. Cow-calf producers in the Northern Plains more often produced small grain crops, mostly wheat, which can be used for cattle grazing in the fall and winter before the summer harvest.

Operator age differed little among beef cow-calf farms in each region, but education was somewhat higher among farm operators in the Southern Plains and West, as a third or more of beef cow-calf producers in these regions completed college. More farm operators had a primary occupation off-farm in the North Central, Southeast, and Southern Plains than in the other regions. Nearly half of farm operators in the Southern Plains, 40 percent of those in the Southeast, and 35 percent in the North Central had a primary occupation off-farm, compared with only 18 percent in the Northern Plains and 23 percent in the West.

Production Practices and Farm Finances by Region

Beef cow-calf technologies and practices were generally used more often by producers in the North Central, Northern Plains, and West regions than those in the Southern Plains and Southeast. For example, producers in these regions used a defined calving season more often than producers in the southern regions (table 5). A defined calving season is common in the North because severe winter weather dictates that cows are scheduled to calve during the spring and calving occurs during a slack labor period for crops commonly grown in the North.

Table 4. Farm and operator characteristics of U.S. beef cow-calf farms,by region, 2008

Item	North Central(a)	Southeast(b)	Northern Plains(c)	Southern Plains(d)	West (e)
Percent of farms/beef cows	16/10	32/24	16/21	27/26	9/18
Beef cows—peak per farm[1]	65^{bcde}	78^{acde}	129^{abde}	99^{abce}	213^{abcd}
Beef cows—average per farm[2]	56^{cde}	59^{cde}	105^{abde}	75^{abce}	155^{abcd}
Beef cows calving	61^{cde}	63^{cde}	118^{abde}	82^{abce}	186^{abcd}
Calves weaned	51^{cde}	51^{cde}	103^{abde}	68^{abce}	154^{abcd}
Weaning age (days)	210^{ce}	206^{ce}	219^{abd}	204^{ce}	222^{abd}
Weaning weight (lbs)	501^{bce}	480^{acde}	543^{abd}	493^{bce}	538^{abd}
Percent of calves:					
Sold at weaning	44^{bde}	70^{ace}	41^{bde}	69^{ace}	53^{abcd}
Backgrounded then sold	45^{bd}	28^{ace}	49^{bde}	29^{ace}	39^{bcd}
Retained until slaughter	11^{bd}	2^{ace}	10^{bd}	2^{ace}	8^{bd}
Total farm production value ($)	$164,037^{bcd}$	$106,779^{acde}$	$241,812^{abde}$	$65,430^{abce}$	$176,869^{bcd}$
Percent from cattle	23^{cde}	25^{cde}	38^{abde}	67^{abc}	66^{abc}
Acres operated	518^{bcde}	453^{acde}	$2,019^{abde}$	$1,436^{abce}$	$4,186^{abcd}$
Private pasture/range acres	208^{bcde}	246^{acde}	$1,359^{abe}$	$1,272^{abe}$	$4,028^{abcd}$
Acres per beef cow	3^{cde}	3^{cde}	11^{abe}	13^{abe}	19^{abcd}
Percent using private pasture/range	94^{bce}	89^{acde}	97^{abe}	97^{be}	82^{abcd}
Percent using public grazing land	0	D	7^{de}	1^{ce}	36^{cde}
Crops produced (percent of farms):					
Corn	37bde	8^{acd}	38^{bde}	2^{abce}	7^{acd}
Soybeans	37^{bd}	3^{ac}	34^{bd}	2^{ac}	0
Small grains	11^{bc}	3^{acde}	43^{abde}	11^{bce}	16^{bcd}
Hay	94^{bcde}	86^{ade}	83^{ade}	62^{abc}	58^{abc}
Operator:					
Age (years)	58^{bd}	61^{ace}	57^{bd}	61^{abe}	59^{bd}
Age (percent greater than age 65)	34	41^{ce}	29^{bd}	39^{ce}	30^{bd}
Completed college (percent)	18^{de}	20^{de}	22^{d}	37^{abc}	33^{abc}
Off-farm occupation (percent)	35^{cde}	40^{cde}	18^{abde}	47^{abce}	23^{abd}
Exit within 5 years (percent)	23	27^{d}	25^{d}	18^{bce}	23^{d}

Source: USDA, Economic Research Service using data from USDA's 2008 Agricultural Resource Management Survey (ARMS).

Notes: D= insufficient data for disclosure. The superscripts refer to the results of statistical tests of differences between item means in each column. All tests are expressed at a 95-percent level of confidence. A lettered superscript denotes that the item mean reported in a column is significantly different from that in the superscript column. The tests were conducted using a jackknife variance estimator with 15 replicates provided with the ARMS data.

[1] Largest number of beef cows and heifers that have calved on farm at any time during 2008.

[2] Average of beginning and ending inventories of beef cows and heifers that have calved on farm during 2008.

A defined calving season also means that the calves are more uniform in size at marketing. Also, producers in the North Central and Northern Plains more often used such technologies as calf implants and/or ionophores (see glossary) than producers in other regions.

Other practices, such as regular veterinary and nutritionist services, were also used more often in the North Central, Northern Plains, and West regions. A majority of beef cow-calf producers in the Southeast and Southern Plains rotated grazing acres, but fewer tested forage quality than in other regions. Information management technologies, such as individual cow records, onfarm computer records, and Internet-based beef cattle information, were generally used less often by producers in the southern regions. For example, 35 percent of beef cow-calf producers in the Southeast kept individual cow records, compared with at least 50 percent of producers in the North Central, Northern Plains, and West regions.

Climatic differences among the regions affect beef cow-calf feed costs. In the northern regions, farms may be required to purchase supplemental forage to sustain beef cows during the winter. In contrast, beef cows in some areas of the South and West regions can graze year round, so less harvested forages are needed to sustain cows during the winter months. Feed costs per cow in 2008 were significantly lower in the Southern Plains, Southeast, and West than in the North Central and Northern Plains (fig. 5). This cost advantage helps account for the relatively larger shares of U.S. beef cows in these regions.

Gross cash income of beef cow-calf farms in 2008 was substantially lower in the Southeast and Southern Plains (about $70,000 per farm) than in the North Central (nearly $160,000 per farm) and in the other regions (over $200,000 per farm) (table 5).

Cattle sales accounted for the majority of cash income in the Southern Plains and West regions. Livestock and crop income were divided more evenly in the Northern Plains and Southeast, and North Central farms generated most of their gross income from crop sales. Net cash and net farm income were highest for beef cow-calf farms in the Northern Plains and North Central, as many farms in these regions benefited from the high prices received for corn and soybeans produced in 2008 (see table 4).

Among the regions, net cash and net farm income were lowest on farms in the Southern Plains and Southeast, where many farm operators worked off-farm to supplement household income. At more than $100,000 per farm, average off-farm income was highest in the Southern Plains, reflecting the high proportion of farm operators working off-farm and the relatively high average education level (see table 4).

Beef cow-calf production accounted for a significant share of the farm household income in the Northern Plains and West regions, as gross cattle sales were nearly $50,000 more than off-farm income.

Gross cattle sales were much lower than off-farm earnings in the North Central, Southeast, and Southern Plains. Average off-farm earnings were higher than the average net income from farming in all regions except the Northern Plains.

The emergence of large dairy and hog farms and the continued shift in production to such farms is well documented (MacDonald et al., 2007; Key and McBride, 2007), but less is known about the economics of large-scale beef cow-calf production. Langemeier et al. (2004) found evidence of economies of size (decreasing costs as size of operation increases) in beef cow-calf production using the National Cattlemen's Association-Integrated Resource Management-Standardized Performance Analysis (NCA-IRM-SPA) database.

Table 5. Production practices and financial characteristics of U.S. beef cow-calf farms, by region, 2008

Item	North Central (a)	Southeast (b)	Northern Plains (c)	Southern Plains (d)	West (e)
Practices:			*Percent using*		
Defined calving season	82^{bcd}	45^{ace}	92^{abde}	42^{ace}	85^{bcd}
Commercial seed stock	16^{c}	13^{c}	21^{abde}	13^{c}	15^{c}
Artificial insemination	11^{bcd}	4^{acde}	17^{abd}	6abce	14^{b}d
Calf implants and/or ionophores	28^{bde}	7^{ace}	26^{bde}	8^{ace}	13^{bd}
Regular veterinary services	27^{bcd}	12^{acde}	41^{abde}	18^{abce}	32^{bcd}
Nutritionist services	11^{bd}	2^{acde}	15^{bde}	5^{abce}	9^{bcd}
Rotational grazing	54^{de}	60^{e}	58^{e}	62^{ae}	71^{abcd}
Tested forage quality	16^{bce}	10^{acde}	27^{abd}	15^{bce}	25^{abd}
Individual cow records	52^{bc}	35^{acde}	59^{abde}	45^{bc}	52^{bc}
Onfarm computer records	20^{e}	13^{cde}	23^{be}	22^{be}	35^{abcd}
Internet for cattle information	33^{e}	29^{cde}	38^{be}	35^{be}	49^{abcd}
Forward purchase inputs	3^{bcde}	$_{7}$ace	10^{abde}	8^{ace}	1$_{7}^{abcd}$
Negotiate input prices	14^{cde}	16^{ce}	21^{abe}	19^{ae}	31^{abcd}
Farm finances:			*Dollars per farm*		
Gross cash income[1]	157,920^{bcd}	73,130^{ace}	228,071^{abd}	66,692^{ace}	203,753^{bcd}
-Cattle sales	37,718^{bcde}	27,012^{acde}	92,407^{abde}	43,691^{abce}	116,223a^{bc} d
-Other livestock sales	13,089	4,314^{cd}	12,016^{bde}	1,980^{bce}	6,674^{de}
-Crop sales	88,526^{bde}	30,582^{acd}	103,011^{bde}	11,167^{abce}	46,419^{acd}
Cash expenses	112,245^{bcd} e	58,097^{acde}	169,367^{abd}	68,160^{abce}	162,149^{abd}
-Variable	79,438^{bcde}	47,975^{ace}	125,594^{abd}	54,644^{ace}	123,197^{abd}
-Fixed	32,807^{bcd}	10,122^{acde}	43,773^{abd}	13,516^{abce}	38,952^{bde}
Net cash farm income	45,676^{bd}	15,033^{acde}	58,704^{bd}	-1,468^{abce}	41,640^{bd}
Net farm income[2]	36,650^{bd}	15,397^{acde}	47,313^{bd}	-7,855^{abce}	33,394^{bd}
Off-farm income[3]	61,311^{cd}	61,385^{cd}	44,952^{abde}	106,266^{abce}	72,010^{cd}

Source: USDA, Economic Research Service using data from USDA's 2008 Agricultural Resource Management Survey (ARMS).

Notes: The superscripts refer to the results of statistical tests of differences between item means in each column. All tests are expressed at a 95-percent level of confidence. A lettered superscript denotes that the item mean reported in a column is significantly different from that in the superscript column. The tests were conducted using a jackknife variance estimator with 15 replicates provided with the ARMS data.

[1] In addition to livestock and crops sales, gross cash income includes Government payments and other farm-related income.

[2] Net farm income equals net cash income less depreciation and the value of noncash benefits for hired labor, plus the net value of commodity inventory change and nonmonetary income.

[3] Off-farm income includes farm household earnings from wages and salaries, earnings from another business, interest and dividend income, pensions and other retirement income, and income from other off-farm sources.

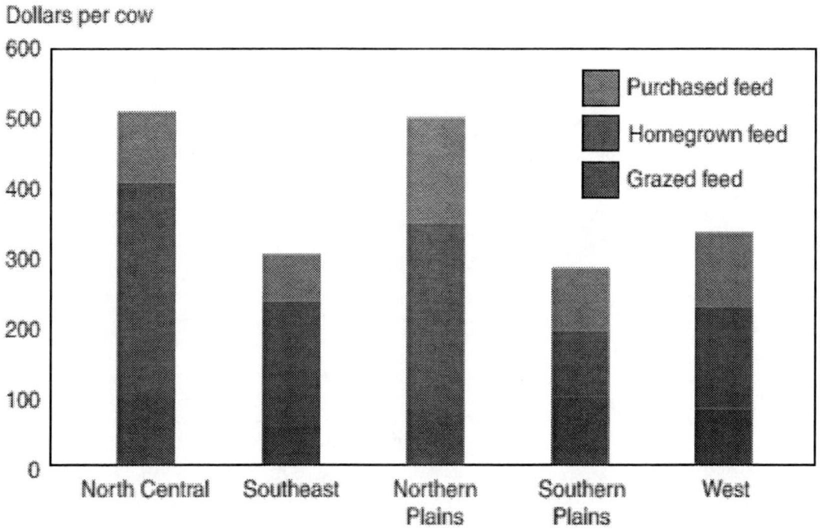

Source: USDA, Economic Research Service using USDA's 2008 Agricultural Resource Management Survey.

Feed costs per cow were lowest in the Southeast, Southern Plains, and West regions because less homegrown forages and other feeds were needed to sustain the beef herd through the winter.

Figure 5. Beef cow-calf feed costs by region, 2008.

SIZE AND COSTS OF BEEF COW-CALF PRODUCTION

The findings indicate that average economic costs per cow for operations declined with farm size up to 500-999 beef cows, but average economic costs per cow were 10 percent higher on operations with 1,000 or more cows than for those with 500-999 cows. The study used data from 206 herds in 20 States that were part of the NCA-IRM-SPA database. Short (2001) also reported finding economies of size in beef cow-calf production using a larger database, but the largest size group included in the analysis was 250 or more cows. Economies of size in beef cow-calf production have also been documented in other studies (Ramsey et al., 2005; Boggs and Hamilton, 1997).

Figure 6 illustrates the relationship between costs of production and size of operation for beef cow-calf operations using data from the 2008 ARMS. Operating, operating plus capital, and total economic costs (see glossary) per cow are highest among the smallest (20-49 cows) producers and lowest among the largest (500 or more cows). Significant economies of size areachieved by moving from the 20-49 cow herd size to the 50-99 cow herd size.[8] Between the 50-99 and 200-499 herd sizes, operating and operating plus capital costs per cow are much the same for the three size groups.

Total economic costs, primarily due to charges for unpaid labor, reveal economies of size across all size groups, and the largest farms (500 cows or more) have significantly lower costs per cow than all other farms.

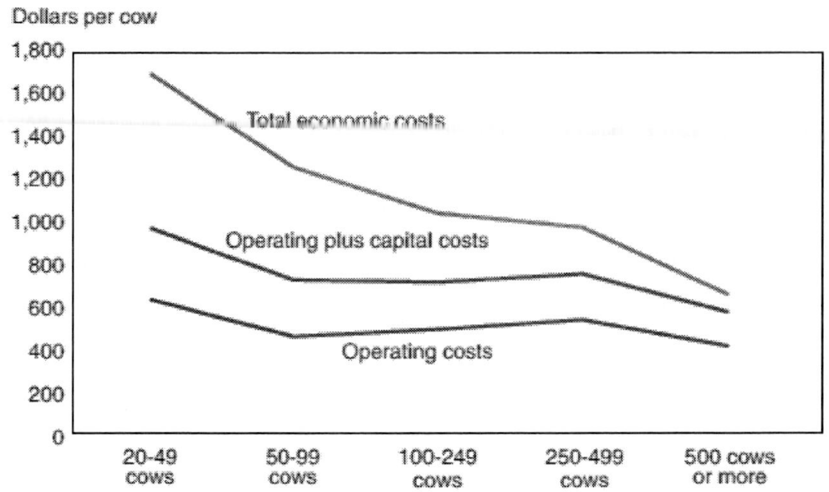

Notes: Production cost estimates for operations with less than 20 beef cows are not available because the ARMS sample is limited to operations with 20 or more beef cows. The number of cows refers to the peak number on the operation at any time during 2008.

Source: USDA, Economic Research Service using USDA's 2008 Agricultural Resource Management Survey (ARMS).

Economies of size are apparent in beef cow-calf production, particularly for total economic costs.

Figure 6. Beef cow-calf cost of production per cow by size, 2008.

Capital and labor costs are much lower on larger operations because they are able to spread fixed units of these resources over greater production. For example, farms with 20-49 cows reported using 31 hours of labor per cow (29 unpaid hours), compared with 6 hours per cow (2.5 unpaid hours) on farms with 500 or more cows. Even if charges for unpaid labor are omitted from production costs, significant economies of size remain.

Despite this evidence for the existence of economies of size in beef cow-calf production, the dramatic shift to larger operations that occurred in hog and milk production was not evident among beef cow-calf farms. Census of Agriculture data reveal that the number of operations with beef cows dropped 15 percent from 1997 to 2007, but the distribution of farms by herd size changed little. In 1997, 82 percent of farms with beef cows had fewer than 50 cows, compared with 80 percent in 2007. Farms with 500 or more cows increased from 0.6 percent of farms with beef cows in 1997 to 0.8 percent in 2007. The beef cow inventory on farms with 500 or more cows increased from 15 percent of the U.S. total in 1997 to 16 percent in 2007 (USDA, NASS, 2007).

Calving and weaning practices used on beef cow-calf operations surveyed in the NAHMS study varied significantly by size of operation. NAHMS data indicate that the largest beef cow-calf operations were more likely to follow a defined calving season than smaller operations.

More than three-fourths of farms in the largest size group (200 or more cows) used spring calving, compared with 38 percent of the smallest operations (1-49 cows) (USDA, APHIS, 2010b). The smallest operations were more likely to sell calves at weaning, whereas the largest were more likely to add value to the calves after weaning by backgrounding (USDA, APHIS, 2010a).

To evaluate the relationship between beef cow-calf farm characteristics, practices, costs, and size of operation, we divided producers surveyed in ARMS into size groups and measured differences in farm structural and financial characteristics among the groups.

Farm and Operator Characteristics by Size

Table 6. Farm and operator characteristics of U.S. beef cow-calf farms, by size of operation (at peak),1 2008

Item	20-49 cows (a)	50-99 cows (b)	100-249 cows (c)	250-499 cows (d)	500 or more cows (e)
Percent of farms/beef cows	41/13	32/21	21/31	5/15	2/20
Beef cows—peak per farm[1]	34bcde	68acde	147^{abde}	332^{abce}	961^{abcd}
Beef cows—average per farm[2]	29bcde	54acde	116^{abde}	260^{abce}	640^{abcd}
Beef cows calving	31bcde	60acde	128^{abde}	292^{abce}	746^{abcd}
Calves weaned	26bcde	49acde	107^{abde}	247^{abce}	622^{abcd}
Weaning age (days)	208^{cde}	209^{cde}	213^{ab}	218^{ab}	214^{ab}
Weaning weight (lbs)	494^{cde}	493^{cde}	523^{abd}	538^{abce}	522^{abd}
Percent of calves:					
Sold at weaning	63^{cde}	62^{cde}	51^{abe}	49^{abe}	39^{abcd}
Backgrounded then sold	31^{cde}	35^{cde}	43^{abe}	45^{abe}	55^{abcd}
Retained until slaughter	6^{b}	3^{acde}	6^{b}	6^{b}	6^{b}
Total farm production value ($)	$68,259^{bcde}$	$93,473^{acde}$	$198,807^{abde}$	$303,144^{abce}$	$931,690^{abcd}$
Percent from cattle	24^{cde}	32^{cde}	43^{abd}	60^{abce}	47^{abd}
Acres operated	346^{bcde}	781^{acde}	$1,654^{abde}$	$5,506^{abce}$	$15,473^{abcd}$
Private pasture/range acres	196^{bcde}	613^{acde}	$1,255^{abde}$	$4,611^{abce}$	$13,134^{abcd}$
Acres per beef cow	6^{bcde}	9^{ae}	9^{ae}	14^{a}	14^{abc}
Percent using private pasture/range	93	92	94	91	90
Percent using public grazing land	1^{bcde}	3^{acde}	8^{abde}	24^{abc}	29^{abc}
Crops produced (percent of farms):					
Corn	15^{bcd}	10^{acd}	25^{abe}	20^{ab}	14^{c}
Soybeans	14^{bcde}	9^{ace}	18^{abde}	9^{ace}	3^{abcd}
Small grains	10^{cde}	11^{cd}	24^{ab}	26^{ab}	19^{a}
Hay	78	76^{c}	81^{be}	78	70^{c}
Operator:					
Age (years)	60^{ce}	61^{cde}	58^{ab}	58^{b}	57^{ab}
Age (percent greater than age 65)	38^{cde}	40^{cde}	32^{abe}	30^{ab}	22^{abc}
Completed college (percent)	23^{cde}	26^{de}	27^{ade}	37^{abc}	42^{abc}
Off-farm occupation (percent)	47^{bcde}	37^{acde}	21^{abe}	18^{ab}	10^{abc}
Exit within 5 years (percent)	26^{cde}	26^{cde}	17^{abde}	10^{abc}	7^{abc}

Source: USDA, Economic Research Service using data from USDA's 2008 Agricultural Resource Management Survey (ARMS).

Notes: The superscripts refer to the results of statistical tests of differences between item means in each column. All tests are expressed at a 95-percent level of confidence. A lettered superscript denotes that the item mean reported in a column is significantly different from that in the superscript column. The tests were conducted using a jackknife variance estimator with 15 replicates provided with the ARMS data.

[1] Largest number of beef cows and heifers that have calved on farm at any time during 2008.

[2] Average of beginning and ending inventories of beef cows and heifers that have calved on farm during 2008.

Beef cow-calf operations were assigned to the following groups based on the reported peak number of beef cows on an operation at any time during 2008: (1) 20-49 cows, (2) 50-99 cows, (3) 100-249 cows, (4) 250-499 cows, and (5) 500 or more cows.

Of the 765,000 U.S. beef cow farms reported in the 2007 Census of Agriculture, about 407,000 had less than 20 cows, whereas more than 200,000 had 20-49 cows. Only 5,800 farms, or less than 1 percent, had 500 or more cows.

Farm specialization in cattle production increased with farm size up to farms with 250-499 beef cows, and then declined among the largest operations. The proportion of farm value of production from cattle peaked at 60 percent among operations with 250-499 beef cows and declined to less than 50 percent for those with 500 or more cows. The largest farm operations contained more than 15,000 acres per farm, with about 13,000 acres of pasture and range land; both of these measurements are nearly three times the average acreage on farms in the next largest size group. Also, 29 percent of farms with 500 or more cows, and 24 percent of those with 250-499 beef cows, used public grazing land, compared with 8 percent or less of farms in the other size groups. The largest size group included many farms in the West and Northern Plains, where public grazing lands are common. More than 60 percent of farms in the smallest size group were in the Southeast and North Central regions (fig. 7).

Farm operator (see glossary) characteristics differed by size among the beef cow-calf producers. Average farm operator age was lower for the largest producers (57 years) than for those with less than 100 cows (60 years), as was the share of farm operators older than age 65 (table 6). Thirty-eight percent of farm operators with less than 100 cows were older than age 65, compared with 22 percent of those with 500 or more cows. The share of farm operators with a primary occupation off-farm declined with farm size, ranging from 47 percent among the smallest producers to 10 percent among the largest. Operator education, indicated by the share graduated from college, increased with size of operation, from 23 percent among the smallest producers to 42 percent among the largest.

The share of producers planning to exit beef cow-calf production in the next 5 years declined with farm size, from 26 percent of those with less than 100 cows to 7 percent of producers with 500 or more cows. This finding suggests that although beef cow-calf production occurs primarily on small operations, the trend is toward increasingly larger operations. However, the number of small operations may not decline as much as suggested if a significant number of retirees purchase farms and begin raising cattle.

The largest beef cow-calf farms more often used advanced production technologies and practices than did smaller farms. Use of a defined calving season, artificial insemination, calf implants and/or ionophores, regular veterinary and nutritionist services, rotational grazing, and forage testing increased with size of operation among beef cow-calf producers (table 7). The use of information management technologies, such as onfarm computer records and Internet-based beef cattle information, also increased with size of operation. The largest producers were also most likely to forward purchase inputs and negotiate input prices. Still, the rate of use of many of these practices was below 50 percent even among the largest producers. On many of the largest farms, the cow-calf operation is a secondary enterprise, and this may account for the low rate of adoption and use of high-tech practices.

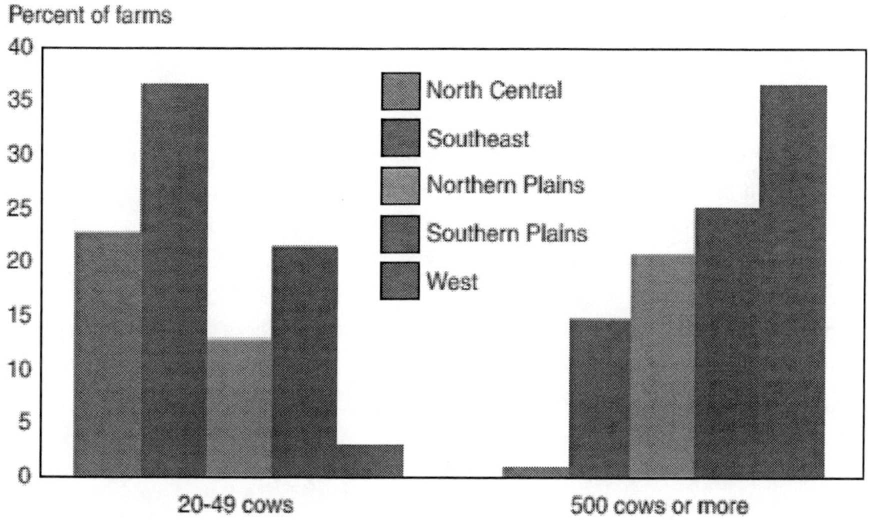

Source: USDA's Economic Research Service using USDA's 2008 Agricultural Resource Management Survey.

Notes: The number of cows refers to the peak number on the operation at any time during 2008. Southeast beef cow-calf farms were most common in the smallest size group, while farms in the West were most common in the largest size group.

Figure 7. The smallest and largest beef cow-calf farms by region, 2008.

Production Practices and Farm Finances by Size

Gross cash income ranged from an average of just over $60,000 per farm on the smallest beef cow-calf operations to nearly $900,000 per farm on the largest. Cattle sales accounted for the largest share of gross cash income on farms in all size groups except the smallest. Cattle sales were most important among operations with 250-499 beef cows, accounting for about 56 percent of gross cash income. Nearly 50 percent of gross cash income on the largest operations was from cattle sales (table 7). The expense structure of beef cow-calf farms exhibits economies of size as fixed cash expenses are spread over more units of production on the largest farms. Fixed cash expenses expressed as a share of total cash expenses decline across the size groups, ranging from 38 percent for producers with 20-49 cows to 18 percent for those with 500 or more cows.

Average cash and net farm income on farms with less than 250 cows are at a level that likely requires many farm households to supplement their income from other sources to reach a subsistence level. Off-farm income, averaging around $70,000 per farm, is likely a critical component of household wellbeing on many of these small farms. Average net farm income was highest for the largest farms, more than three times that of farms with 250-499 beef cows. Households of the largest farms' operators also generated an average off-farm income that was nearly twice that of the other size groups.

ARMS data indicate that over a third (36 percent) of beef cow-calf producers cite some type of off-farm employment as their main occupation. This share is much higher than that for other types of livestock and crop producers and suggests that cow-calf producers have a diverse set of goals for their farm operations.

A Farm Typology of Beef Cow-Calf Producers

Genter and Tanka (2002) identified eight clusters, or distinct sets, of cattle and sheep ranchers based on their attributes and attitudes about the farm business. Their study found ranchers to be very heterogeneous, with many placing more value in ranching as a tradition and a place to raise a family than as a profit-generating endeavor.

NAHMS data provide other information about the importance of beef cow-calf production to household income. Only 5 percent of farms with 50 or fewer cows reported that the beef cow-calf operation was a primary source of income (USDA, APHIS, 2008). This percentage increased with size of operation to where 65 percent of operations with 200 or more beef cows regarded the beef cow-calf operation as a primary source of income. Among regions, the West had the highest share (25 percent) of farms that considered the beef cow-calf operation to be a primary income source. Overall, NAHMS data show that only 14 percent of U.S. farms with beef cows regarded beef cow-calf production as a primary source of income.

We divided beef cow-calf producers in ARMS into three groups based on a farm typology that reflects different producer characteristics and goals (Hoppe et al., 2000). The first group is *rural residence farms*, those with annual gross sales below $250,000 and operators who consider farming to be a secondary activity in terms of resources invested in the farm and the amount of income it contributes to the farm household. Rural residence farms comprise three groups: (1) limited resource farms, those with sales less than $100,000, farm assets less than $150,000, and total household income less than $20,000; (2) farms whose operators report that they are retired; and (3) residential lifestyle farms, whose operators report a major occupation off-farm. The second typology group is *intermediate farms*, those with annual sales below $250,000 and operators who report farming as their main occupation. The final group, *commercial farms*, consists of large family farms with annual sales above $250,000 and some nonfamily enterprises organized as cooperatives or nonfamily corporations, and farms with a hired manager.

Farm and Operator Characteristics by Farm Typology

Half of beef cow-calf producers were classified as rural residence farms, and these farms were smaller than other beef cow-calf farms, accounting for about a third of total beef cows (table 8). Only 13 percent of beef cow-calf producers were commercial farms, and these farms accounted for 30 percent of beef cows. Rural residence farms averaged 66 beef cows (at peak), compared with 101 beef cows (at peak) on intermediate farms and 244 beef cows (at peak) on commercials farms. Two-thirds of the calves produced on rural residence farms were sold at weaning, compared with 36 percent on commercial farms. Commercial farms were more likely to retain calves for backgrounding (48 percent) and for finishing (16 percent).

The value of farm production and the importance of cattle on the beef cow-calf farms differed greatly among the farm typology groups. Commercial farms generated on average $700,000 of farm product value in 2008, compared with about $75,000 on intermediate farms and $32,000 on rural residence farms. Cattle accounted for 27 percent of the production value on commercial farms, compared with 57 percent on intermediate farms and 75 percent on rural residence farms.

Table 7. Production practices and financial characteristics of U.S. beef cow-calf farms, by size of operation (at peak), 2008

Item	20-49 cows (a)	50-99 cows (b)	100-249 cows (c)	250-499 cows (d)	500 or more cows (e)
Practices:			*Percent using*		
Defined calving season	55^{cde}	57^{cde}	73^{abe}	78^{ab}	82^{abc}
Commercial seed stock	15^{e}	15^{e}	15^{e}	17e	24^{abcd}
Artificial insemination	$_5$cde	$_7$cde	13^{abde}	19^{abce}	24^{abcd}
Calf implants and/or ionophores	12^{cde}	12^{cde}	19^{abe}	21abe	26^{abcd}
Regular veterinary services	16^{cde}	20^{cde}	34^{abe}	41^{ab}	43^{abc}
Nutritionist services	6^{bcde}	2^{acde}	13^{abde}	18^{abc}	25^{abc}
Rotational grazing	54^{bcde}	62^{ade}	66^{ade}	75^{abc}	71^{abc}
Tested forage quality	10^{bcde}	15^{acde}	25^{abe}	32^{ab}	39^{abc}
Individual cow records	41^{cd}	44^{cd}	55^{ab}	58^{ab}	50^{ab}
Onfarm computer records	$12b^{cde}$	22^{acde}	27^{abde}	36^{abce}	49^{abcd}
Internet for cattle information	27^{bcde}	34^{acd}e	45^{ade}	46^{ace}	48^{abcd}
Forward purchase inputs	4^{bcde}	6^{acde}	12^{abde}	22^{abce}	38^{abcd}
Negotiate input prices	11bcde	20^{ade}	24^{ade}	43^{abc}	53^{abc}
Farm finances:			*Dollars per farm*		
Gross cash income[1]	$60,443^{bcde}$	$76,775^{acde}$	$185,598^{abde}$	$324,910^{abce}$	$883,310^{abcd}$
-Cattle sales	$16,522^{bcde}$	$29,675^{acde}$	$85,831^{abde}$	$180,530^{abce}$	$439,474^{abcd}$
-Other livestock sales	$1,525^{cde}$	$4,773^{de}$	$7,320^{ade}$	$36,082^{abc}$	$56,047^{abc}$
-Crop sales	$32,767^{cde}$	$27,764^{cde}$	$69,577^{abe}$	$74,802^{abc}$	$355,180^{abcd}$
Cash expenses	$46,924^{bcde}$	$66,157^{acde}$	$146,561^{abde}$	$245,539^{abce}$	$689,617^{abcd}$
-Variable	$34,022^{bcde}$	$50,089^{acde}$	$112,460^{abde}$	$189,215^{abce}$	$568,869^{abcd}$
-Fixed	$12,903^{bcde}$	$16,067^{acde}$	$34,101^{abde}$	$56,325^{abce}$	$120,693^{abcd}$
Net cash farm income	$13,519^{cde}$	$10,619^{cde}$	$39,037^{abde}$	$79,370^{abce}$	$193,693^{abcd}$
Net farm income[2]	$11,410^{cde}$	$9,253^{cde}$	25,508abde	$55,484^{abce}$	$175,935^{abcd}$
Off-farm income[3]	$71,487^{e}$	$68,301^{e}$	$70,613^{e}$	$78,421^{e}$	$144,619^{abcd}$

Source: USDA, Economic Research Service using data from USDA's 2008 Agricultural Resource Management Survey (ARMS).

Note: The superscripts refer to the results of statistical tests of differences between item means in each column. All tests are expressed at a 95-percent level of confidence. A lettered superscript denotes that the item mean reported in a column is significantly different from that in the superscript column. The tests were conducted using a jackknife variance estimator with 15 replicates provided with the ARMS data.

[1] In addition to livestock and crops sales, gross cash income includes Government payments and other farm-related income.

[2] Net farm income equals net cash income less depreciation and the value of noncash benefits for hired labor, plus the net value of commodity inventory change and nonmonetary income.

[3] Off-farm income includes farm household earnings from wages and salaries, earnings from another business, interest and dividend income, pensions and other retirement income, and income from other off-farm sources.

Table 8. Farm and operator characteristics of U.S. beef cow-calf farms, by farm typology, 2008

Item	Rural residence farms (a)	Intermediate farms (b)	Commercial farms (c)
Percent of farms/beef cows	50/32	38/38	13/30
Beef cows—peak per farm[1]	66[bc]	101[ac]	244[ab]
Beef cows—average per farm[2]	51[bc]	79[ac]	186[ab]
Beef cows calving	57[bc]	88[ac]	211[ab]
Calves weaned	46[bc]	73[ac]	181[ab]
Weaning age (days)	206[bc]	213[a]	215[a]
Weaning weight (lbs)	491[bc]	511[ac]	523[ab]
Percent of calves:			
Sold at weaning	67[bc]	57[ac]	36[ab]
Backgrounded then sold	31[bc]	37[ac]	48[ab]
Retained until slaughter	2bc	6ac	16ab
Total farm production value ($)	31,559[bc]	74,561[ac]	708,462[ab]
Percent from cattle	75[bc]	57[ac]	27[ab]
Acres operated	720[bc]	1,135[ac]	4,217[ab]
Private pasture/range acres	622[bc]	941[ac]	2,832[ab]
Acres per beef cow	9	9[c]	12[b]
Percent using private pasture/range	94[b]	90[ac]	93[a]
Percent using public grazing land	2bc	7ac	9ab
Crops produced (percent of farms) Corn	6[bc]	16[ac]	55[ab]
Soybeans	4[bc]	14[ac]	45[ab]
Small grains	6[bc]	17[ac]	39[ab]
Hay	76[c]	79	83[a]
Operator:			
Age (years)	60[bc]	62[ac]	55[ab]
Age (percent greater than age 65)	36[bc]	43[ac]	19[ab]
Completed college (percent)	28[b]	23[a]	27
Off-farm occupation (percent)	71[c]	0	6[a]
Exit within 5 years (percent)	25[c]	24[c]	14[ab]

Source: USDA, Economic Research Service using data from USDA's 2008 Agricultural Resource Management Survey (ARMS).

Notes: The superscripts refer to the results of statistical tests of differences between item means in each column. All tests are expressed at a 95-percent level of confidence. A lettered superscript denotes that the item mean reported in a column is significantly different from that in the superscript column. The tests were conducted using a jackknife variance estimator with 15 replicates provided with the ARMS data.

[1] Largest number of beef cows and heifers that have calved on farm at any time during 2008.

[2] Average of beginning and ending inventories of beef cows and heifers that have calved on farm during 2008.

Among beef cow-calf operations, shares of rural residence and commercial farms varied by region. Nearly 70 percent of rural residence farms were located in the Southeast and Southern Plains, with each region accounting for about a third (fig. 8). In contrast, 60 percent of commercial beef cow-calf farms were located in the Northern Plains and West regions. Operator characteristics of beef cow-calf farms also varied significantly by farm typology.

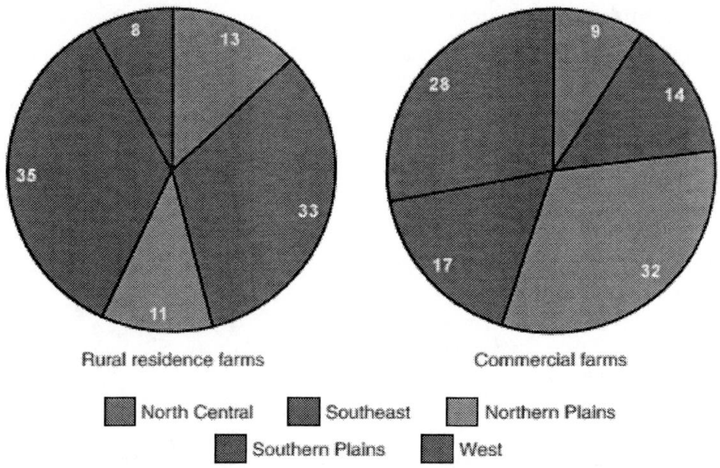

Source: USDA's Economic Research Service using USDA's 2008 Agricultural Resource Management Survey.

Southeast and Southern Plains beef cow-calf farms accounted for nearly 70 percent of rural residence farms, while most commercial farms were in the Northern Plains and West.

Figure 8. Rural residence and commercial beef cow-calf farms by region, 2008.

Operators of rural residence and intermediate farms, on average, were 5-7 years older than operators of commercial farms, and more were over age 65. Thirty-six percent of rural residence farm operators and 43 percent of intermediate farm operators were over age 65, compared with only 19 percent of commercial farm operators. By definition, more rural residence farm operators (71 percent) had a major occupation off-farm than operators of other farms.

Also, about a quarter of rural residence and intermediate farm operators expected to exit cow-calf production in the next 5 years, compared with 14 percent of commercial farm operators.

Production Practices and Farm Finances by Farm Typology

Commercial beef cow-calf farms were more likely than other farm types to use advanced technologies and practices, such as a defined calving season, calf implants and/or ionophores, veterinary and nutritionist services, and forage quality testing (table 9). Commercial farms were also more likely to use information management technologies, such as onfarm computer records and the Internet, and input purchasing practices, such as forward purchasing and negotiating discounts (see glossary).

The average gross cash income in 2008 on commercial farms was over $600,000, compared with $77,000 on intermediate farms and $34,000 on rural residence farms. The composition of gross cash income on farms in each typology group reflects their relative specialization. Most of gross cash income on rural residence and intermediate farms was from cattle, while commercial farms generated most of gross cash income from crops. Net cash and net farm incomes in 2008 were about $174,000 and $136,000, respectively, on commercial farms. In contrast, intermediate farms earned about $11,000 and $7,000 in net cash and net farm income.

Average net cash and farm incomes were negative in 2008 on rural residence beef cow-calf farms. Operators of rural residence farms often have goals other than profitability, as most work off-farm or are retired, and farming may be regarded as a lifestyle choice.

Off-farm earnings on rural residence farms averaged nearly $90,000 per farm in 2008. In contrast, running a profitable enterprise is important to operators of commercial farms, where net farm income was significantly higher than off-farm income. Operators of intermediate farms, by definition, cite farming as their main occupation. Low farm income and relatively low off-farm income on these farms suggest that they are more vulnerable to changing economic conditions.

Beef Cow-Calf Production and the National Animal Identification System

Historically, U.S. beef cattle producers have had three primary motives for establishing a traceability system for live animals: (1) protect property from theft or loss, (2) control the spread of animal diseases, and (3) provide proof of credence attributes that add product value (Golan et al., 2004). During the past 20 years, however, protecting consumer confidence in beef products has emerged as an additional motive. This led to a series of plans that resulted in the implementation of the National Animal Identification System (NAIS) in 2004 (see box, “Evolution of the National Animal Identification System”).

The foundation of the NAIS and the first step for producer participation was premises registration.

To register a premises, producers contacted the appropriate State (e.g., State Veterinarian’s office) or tribal authority and provided contact information for the premises and minimal information about the livestock operation (USDA, APHIS, 2007). Producers who provided premises information were to be notified quickly when a disease event might affect their area or animal species. The intention of premises registration was to enable animal health officials to quickly locate at-risk animals and take actions to address emergency situations, minimize hardships, and speed disease eradication efforts.

After registering a premises, a producer could participate in the animal identification component of the NAIS. Animals of the same species that typically move through the production chain as a group could be identified by a group or lot identification number (GIN), determined by the animal owner using the premises identification number and the date the group was assembled. Animals that move through the chain individually had to be identified with a USDA-recognized animal identification number (AIN) tag or device. Once producers identified their animals, they could choose to report animal records to an animal tracking device (ATD) of their choice (USDA, APHIS, 2007).

In contrast to other U.S. livestock sectors, there was resistance to the NAIS among some beef cow-calf producers. A study of beef cow-calf producer preferences for voluntary traceability systems found that producers are sensitive to cost, the managing entity, and information requirements of the system (Schulz and Tonsor, 2010). Costs of the NAIS could be significant for small operations because beef cattle have to be individually identified. A USDA study of NAIS costs for full traceability reported that cattle industry costs represented 91.5 percent of the total annual costs of the NAIS for the primary food animal species (cattle, sheep, swine, and poultry). The study estimated the total cost of the NAIS for cattle producers in aggregate as $175.9 million annually at a 90-percent participation level (USDA, APHIS, 2009).

Table 9. Production practices and financial characteristics of U.S. beef cow-calf farms, by farm typology, 2008

Item	Rural residence farms (a)	Intermediate farms (b)	Commercial farms (c)
Practices:		*Percent using*	
Defined calving season	52[bc]	77[ac]	81[ab]
Commercial seed stock	14[c]	14[c]	19[ab]
Artificial insemination	1	3	2
Calf implants and/or ionophores	10[c]	13[c]	33[ab]
Regular veterinary services	18[bc]	25[ac]	35[ab]
Nutritionist services	4bc	8ac	18ab
Rotational grazing	60	52	59
Tested forage quality	13[bc]	16[ac]	31[ab]
Individual cow records	44[c]	47	51[a]
Onfarm computer records	20[c]	18[c]	29[ab]
Internet for cattle information	32[c]	33[c]	51[ab]
Forward purchase inputs	7c	7c	17ab
Negotiate input prices	14[bc]	21[ac]	33[ab]
Farm finances:		Dollars per farm	
Gross cash income[1]	33,677[bc]	77,165[ac]	602,849[ab]
-Cattle sales	23,720[bc]	42,202[ac]	191,789[ab]
-Other livestock sales	729[bc]	1,299[ac]	45,186[ab]
-Crop sales	4,618[bc]	18,493[ac]	305,683[ab]
Cash expenses	36,212[bc]	66,142[ac]	428,210[ab]
-Variable	26,738[bc]	49,555[ac]	335,106[ab]
-Fixed	9,474[bc]	16,587[ac]	93,104[ab]
Net cash farm income	-2,535[bc]	11,023[ac]	174,639[ab]
Net farm income[2]	-803[bc]	6,669[ac]	136,014[ab]
Off-farm income[3]	89,929[bc]	53,571[a]	53,991[a]

Source: USDA, Economic Research Service using data from USDA's 2008 Agricultural Resource Management Survey (ARMS).

Notes: The superscripts refer to the results of statistical tests of differences between item means in each column. All tests are expressed at a 95-percent level of confidence. A lettered superscript denotes that the item mean reported in a column is significantly different from that in the superscript column. The tests were conducted using a jackknife variance estimator with 15 replicates provided with the ARMS data.

[1]In addition to livestock and crops sales, gross cash income includes Government payments and other farm-related income. [2]Net farm income equals net cash income less depreciation and the value of noncash benefits for hired labor, plus the net value of commodity inventory change and nonmonetary income. [3]Off-farm income includes farm household earnings from wages and salaries, earnings from another business, interest and dividend income, pensions and other retirement income, and income from other off-farm sources.

The USDA study also estimated costs of full traceability for the beef cow-calf sector by size of operation and separately for those who currently tag animals and those that do not.

Economies of size were found to exist for full participation in the NAIS, as costs ranged from $2.48 per cow for the largest operations (5,000 or more cows) currently tagging cattle to $6.16 per cow for the smallest operations (1-49 cows) not currently tagging cattle.[9] Most economies of size were captured by operations with 50-99 cows, compared with those with fewer cows.

Evolution of the National Animal Identification System

The importance of having an animal identification system was highlighted during 1986-88 when bovine spongiform encephalopathy (BSE), or mad cow disease—a fatal neurological disease—was identified in cattle herds of the United Kingdom. Subsequent testing found BSE to be widespread among the UK cattle population, resulting in the slaughter of 3.7 million head. In 1997, an outbreak of foot-and-mouth disease in swine in Taiwan caused $6.9 billion in losses and eradication costs, including the slaughter of 3.8 million pigs, and decimated Taiwan's previously strong pork export market (Schnepf, 2010).

In May 2003, Canadian officials announced that a single cow from a farm in Alberta was diagnosed as having BSE. All beef and cattle exports from Canada immediately stopped, devastating the Canadian beef industry, where approximately half of annual production is exported (Lawrence et al., 2003).

To ensure confidence in the U.S. beef supply, among the initiatives USDA announced was the acceleration of a verifiable national animal identification system. The U.S. Animal Identification Plan (USAIP) called for the establishment of individual premises identification by the summer of 2004, individual animal identification by 2005, and full implementation and compliance (all covered species and their movements—both interstate and intrastate) by July 2006 (Schnepf, 2010). The USAIP evolved into the National Animal Identification System (NAIS), retaining most of its essential elements.

Participation in the NAIS was originally planned to be mandatory, but in response to various concerns raised by livestock producers and small farmers, USDA announced in August 2006 that the NAIS would be entirely voluntary at the Federal level. During 2006-09, USDA released a series of draft reports that developed the framework of the NAIS. In 2009, the U.S. Secretary of Agriculture undertook a series of listening sessions with stakeholders from around the country to gather feedback and help shape decisions about the future direction of animal identification and traceability in the United States.

In February 2010, USDA announced substantial revisions in its approach to achieving a national capacity for animal disease traceability. The NAIS was replaced with a new approach that allows individual States (and tribal nations) to choose their own degree of within-State animal identification and traceability of livestock. USDA did require that all animals moving in interstate commerce have a form of identification that allows traceability back to the originating State (Schnepf, 2010).

For example, among operations currently tagging, costs declined $1.82 per cow between the 1-49 and the 50-99 cow groups and only an additional $0.82 between the 50-99 and 5,000 or more cow groups. However, the high costs faced by the smallest operations would affect most U.S. cow-calf producers, as nearly 80 percent have fewer than 50 cows.

Familiarity and Registration with the NAIS

Beef cow-calf producers surveyed in ARMS were asked if they were familiar with NAIS, and if so, were their premises registered with the NAIS.[10] Producers were classified into one of three groups: (1) those not familiar with NAIS, (2) those familiar, but not registered with NAIS, and (3) those registered with NAIS. Among beef cow-calf producers, 24 percent reported not being familiar with NAIS, 52 percent were familiar but had not registered their premises with NAIS, and 24 percent had registered their premises with NAIS (table 10). This compares with the 16 percent of beef cow-calf operations that had a unique premises registration found in the 2007 NAHMS study (USDA, APHIS, 2008).[11]

There was an association between familiarity and registration with NAIS and size of cow-calf operation. The 24 percent of farms registered with NAIS accounted for a third of beef cows, while the 24 percent not familiar with NAIS accounted for only 16 percent. Operations registered with NAIS had an average of 140 beef cows (at peak), double the average size of those not familiar with NAIS. Nearly a third of cow-calf operations with 20-49 beef cows were not familiar with the NAIS, and only 16 percent of these small operations were registered. In contrast, only 7 percent of operations with 500 or more cows were not familiar with NAIS, and nearly half of these farms were registered (fig. 9).

NAIS registration was associated with location, as registration rates were below 20 percent in many Southeast States (e.g., AL, GA, MS, FL, and KY) but above 40 percent in States in the West (e.g., CA and CO). Differences in farm operator characteristics were also significant among the groups. Cow-calf producers not familiar with NAIS were 5 years older, on average, than other producers, and nearly half were over age 65 (table 10). Producers registered with NAIS were more educated than other producers, as a third had graduated from college, compared with 17 percent not familiar with NAIS and 27 percent familiar but not registered. Also, a higher percentage of cow-calf producers not familiar with NAIS planned to exit the beef cow-calf business within 5 years.

Cow-calf producers registered with NAIS were more likely to use advanced technologies and practices than other producers. Those registered were more likely to have a defined calving season, artificially inseminate cows, use growth-promoting implants and/or ionophores, use regular veterinary and nutritionist services, and manage forage quality by rotational grazing and forage testing (table 11).

Differences in the use of information technologies were also significant among beef cow-calf producers in each group. More than 60 percent of those registered in the NAIS kept individual cow records, 30 percent kept onfarm computer records for beef cattle, and about half used the Internet to obtain beef cattle information, all much higher usage rates than among other cow-calf producers.

More than 80 percent of beef cow-calfproducers used some method for identifying cattle, including more than 90 percent of those with premises registered with NAIS and nearly 70 percent of those not familiar with NAIS (fig. 10). Visual ear tagging and branding were the most common methods used to identify beef cattle.

The average income per farm of cow-calf producers registered with NAIS was much higher than those of other cow-calf producers. Net cash farm income and net farm income were more than three times higher than for those not familiar with NAIS and $8,000-$10,000 more than that for those familiar with but not registered (table 11). Off-farm incomes were similar among the groups, from about $67,000 to $73,000, all much higher than net farm income.

Table 10. Farm and operator characteristics of U.S. beef cow-calf operations, by producer familiarity and registration with the NAIS, 2008

Item	Not familiar (a)	Familiar, not registered (b)	Registered (c)
Percent of farms/beef cows	24/16	52/50	24/32
Beef cows—peak per farm[1]	70^{bc}	98^{ac}	140^{ab}
Beef cows—average per farm[2]	56^{bc}	77^{ac}	106^{ab}
Beef cows calving	62^{bc}	85^{ac}	120^{ab}
Calves weaned	51^{bc}	71^{ac}	101^{ab}
Weaning age (days)	209	211	208
Weaning weight (lbs)	486^{bc}	504^{ac}	515^{ab}
Percent of calves:			
Sold at weaning	62^{c}	61^{c}	52^{ab}
Backgrounded then sold	34^{c}	34^{c}	41^{ab}
Retained until slaughter	4c	5c	7ab
Total farm production value ($)	84,178^{bc}	130,391^{ac}	185,446^{ab}
Percent from cattle	34	39	41
Acres operated	759^{bc}	1,242^{ac}	1,824^{ab}
Private pasture/range acres	565^{bc}	962^{ac}	1,391^{ab}
Acres per beef cow	8^{c}	10	10^{a}
Percent using private pasture/range	91	94^{c}	91^{b}
Percent using public grazing land	4	5	5
Crops produced (percent of farms) Hay	76	79	79
Small grains	12	14	16
Location (percent of farms):			
North Central	14	18^{c}	14^{b}
Southeast	37^{bc}	30^{a}	30^{a}
Northern Plains	14	17	17
Southern Plains	29	26	29
West	7bc	9a	10a
Operator:			
Age (years)	63^{bc}	59^{a}	58^{a}
Age (percent greater than age 65)	47^{bc}	34^{a}	32^{a}
Off-farm occupation (percent)	39	36	34
Completed college (percent)	17^{bc}	27^{ac}	33^{ab}
Exit within 5 years (percent)	31^{bc}	21^{a}	20^{a}

Source: USDA, Economic Research Service using data from USDA's 2008 Agricultural Resource Management Survey (ARMS).

Notes: NAIS = National Animal Identification System. The superscripts refer to the results of statistical tests of differences between item means in each column. All tests are expressed at a 95-percent level of confidence. A lettered superscript denotes that the item mean reported in a column is significantly different from that in the superscript column. The tests were conducted using a jackknife variance estimator with 15 replicates provided with the ARMS data.

[1] Largest number of beef cows and heifers that have calved on farm at any time during 2008.

[2] Average of beginning and ending inventories of beef cows and heifers that have calved on farm during 2008.

Table 11. Production practices and financial characteristics of U.S. beef cow-calf operations, by producer familiarity and registration with the NAIS, 2008

Item	Not familiar (a)	Familiar, not registered (b)	Registered (c)
Practices:		*Percent using*	
Defined calving season	50^{bc}	63^{ac}	70^{ab}
Commercial seed stock	12	16	16
Artificial insemination	2bc	8ac	16ab
Calf implants and/or ionophores	12^{c}	12^{c}	22^{ab}
Regular veterinary services	17^{bc}	21^{ac}	32^{ab}
Nutritionist services	3bc	7ac	12ab
Rotational grazing	54^{bc}	62^{a}	65^{a}
Tested forage quality	8^{bc}	17^{ac}	24^{ab}
Individual cow records	31^{bc}	46^{ac}	63^{ab}
Onfarm computer records	10^{bc}	20^{ac}	32^{ab}
Internet for cattle information	18^{bc}	35^{ac}	51^{ab}
Forward purchase inputs	5bc	7ac	14ab
Negotiate input prices	12^{bc}	19^{ac}	27^{ab}
Farm finances:		*Dollars per farm*	
Gross cash income[1]	$70,356^{bc}$	$116,965^{ac}$	$180,276^{ab}$
-Cattle sales	28,385bc	50,602ac	76,493ab
-Livestock sales	$1,712^{bc}$	$4,702^{ac}$	$14,466^{ab}$
-Crop sales	$28,993^{bc}$	$44,925^{ac}$	$72,594^{ab}$
Cash expenses	$59,222^{bc}$	$90,885^{ac}$	$143,872^{ab}$
-Variable	$43,977^{bc}$	$68,569^{ac}$	$113,447^{ab}$
-Fixed	$15,245^{bc}$	$22,315^{ac}$	$30,425^{ab}$
Net cash farm income	$11,134^{bc}$	$26,081^{ac}$	$36,404^{ab}$
Net farm income[2]	$7,517^{bc}$	$20,407^{a}$	$28,772^{a}$
Off-farm income[3]	66,607	73,104	72,648

Source: USDA, Economic Research Service using data from USDA's 2008 Agricultural Resource Management Survey (ARMS).

Notes: NAIS = National Animal Identification System. The superscripts refer to the results of statistical tests of differences between item means in each column. All tests are expressed at a 95-percent level of confidence. A lettered superscript denotes that the item mean reported in a column is significantly different from that in the superscript column. The tests were conducted using a jackknife variance estimator with 15 replicates provided with the ARMS data.

[1] In addition to livestock and crops sales, gross cash income includes Government payments and other farm-related income.

[2] Net farm income equals net cash income less depreciation and the value of noncash benefits for hired labor, plus the net value of commodity inventory change and nonmonetary income.

[3] Off-farm income includes farm household earnings from wages and salaries, earnings from another business, interest and dividend income, pensions and other retirement income, and income from other off-farm sources.

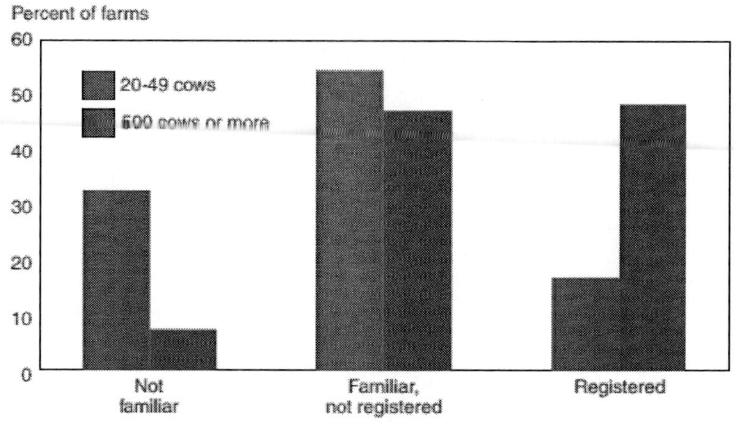

Source: USDA, Economic Research Service using USDA's 2008 Agricultural Resource Management Survey.

NAIS = National Animal Identification System.

Only 16 percent of beef cow-calf producers with 20-49 cows were registered with the NAIS, compared with nearly half of those with 500 cows or more.

Figure 9. NAIS familiarity and registration by size, 2008.

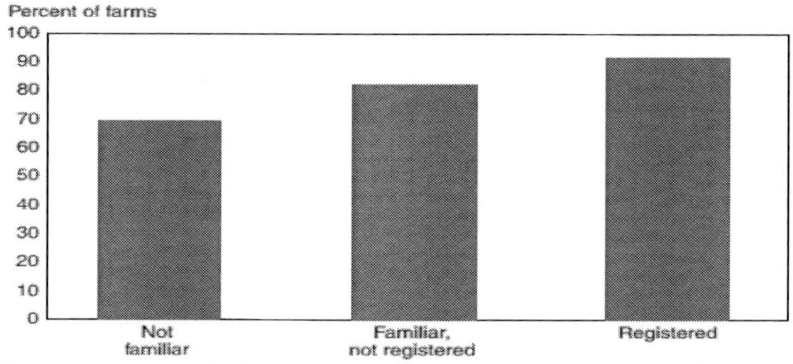

Source: USDA, Economic Research Service, using USDA's 2008 Agricultural Resource Management Survey.

NAIS = National Animal Identification System.

Most beef cow-calf producers used some type of cattle ID (e.g., ear tagging, branding) regardless of NAIS participation.

Figure 10. Use of cattle ID by NAIS familiarity and registration, 2008.

Conclusion

Analysis of 2008 ARMS data reveals a commercial beef cow-calf industry in the United States that is diverse in type, size, location, and characteristics of farm operators. About half of beef cow-calf farms produce calves that are sold at or shortly after weaning. These are usually small farms, and most are located in the Southeast and Southern Plains. Many of these farm households rely on off-farm income to supplement a low level of farm income. Most of the other beef cow-calf farms retain ownership of calves after weaning and continue grazing, or backgrounding, the calves, from 30 to 90 days before selling the calves to feedlots. These farms are generally larger, have more beef cows, and are distributed throughout the United States, with many in the Northern Plains and West regions. A small percentage of beef cow-calf producers retain ownership of the calves after weaning until the animals reach slaughter weight. These operations are primarily located in areas where corn is produced for livestock feed, such as the North Central and Northern Plains.

The majority of U.S. beef cows are located in the South, including the Southern Plains (primarily Texas) and the Southeast. These regions have the advantage of a longer grazing season, meaning that less supplemental forage is usually required to support beef cattle during the winter, resulting in lower feed costs.

Despite higher feed costs in the Northern Plains, large beef cow-calf producers in this region have been able to compete with those in the South due to efficient production methods and economies of size.

Economies of size in beef cow-calf production suggest that farms have an incentive to become larger. However, land area required for large-scale beef cow-calf production makes it difficult for beef cow-calf producers to take advantage of economies of size. In most areas of the United States, beef cow-calf production is the residual user of land. As the opportunity cost of pasture and range lands increases for such uses as crop production or recreational activities, the size of beef cow-calf operations may be limited or fragmented into smaller units.

Unlike farms in other livestock industries, most farms with beef cows do not specialize in beef cow-calf production. For example, on hog and dairy farms, the average share of farm product value from hog and milk production is more than 70 and 80 percent, respectively (McBride and Key, 2007; Short, 2001). On farms with beef cows, less than 40 percent of farm product value stems from cattle production.[12] The degree of specialization in beef cow-calf production varies significantly by region. About two-thirds or more of farm product value on farms in the Southern Plains and West regions was from cattle production, compared with less than 40 percent of farms in other regions. Size of operation as measured by the number of beef cows did not necessarily indicate specialization in beef cow-calf production.

Operations with 250-499 beef cows earned 60 percent of farm product value from cattle; the share dropped to less than 50 percent for operations with 500 or more cows.

Beef cow-calf operations also differed significantly from other livestock operations in the extent to which farm operators have a primary occupation off-farm. More than a third of beef cow-calf farm operators worked off-farm, and half of the farms were classified as rural residence-farms. Less than 10 percent of hog and dairy farms are classified as rural residence farms (McBride and Key, 2007; Short, 2001).These rural residence farms were small operations that specialized in beef cow-calf production (75 percent of farm product value from cattle), but off-farm income was their primary household income source.

Commercial farms, those with annual sales of at least $250,000, with beef cow-calf enterprises were mostly diversified operations on which beef cattle was a secondary enterprise, accounting for about a fourth of farm product value. Intermediate farms, those with annual farm sales less than $250,000 and with operators whose main occupation is farming, depended more on the beef cattle enterprise as a farm income source, with more than half of farm product value coming from cattle.

The low levels of net cash and net farm income generated by these intermediate farms in 2008 suggest that farms in this group are among the most financially vulnerable to the input and output price variations of beef cattle production.

In 2008, more than 80 percent of beef cow-calf producers had some type of animal identification system in place, such as branding or ear tagging. But, nearly a quarter of beef cow-calf producers with 20 or more cows reported a lack of familiarity with the NAIS, and only about a quarter had their premises registered with the system.

This lack of participation among the nearly 765,000 U.S. beef cow-calf producers appears to be related to concerns about liability and costs associated with the program. Since beef cow-calf production is a secondary farm enterprise and a secondary household income source for most farms with beef cows, there may be little incentive for these farms to risk any perceived liability or to incur program participation costs.

This creates a challenge for Federal or State efforts to enhance traceability through animal identification on beef cow-calf farms.

Glossary

Artificial insemination is the process by which sperm is placed in the reproductive tract of a beef cow or heifer for the purpose of impregnating the cow or heifer by means other than natural service from a bull.

Backgrounding is an intermediate stage sometimes used in beef cattle production that begins after weaning calves and ends when calves are placed in a feedlot to be fed to a slaughter weight. Feeding during the backgrounding phase relies more heavily on forages (e.g., pasture, hay, crop residues) in combination with grains to increase a calf's weight by several hundred pounds before it enters a feedlot (see stocker calves). Backgrounding is often conducted on farms where calves are born, but some farms specialize in backgrounding cattle.

Beef cow-calf farms are represented by those selected in a targeted sample of beef cow-calf farms as part of USDA's 2008 Agricultural Resource Management Survey. Beef cow-calf farms are defined as farms that had a beef cow inventory of 20 or more head on the acres operated at any time during 2008. This excludes a large portion of farms with beef cows but covers most of the beef cows.

Commercial seed stock producers are cattle producers who specialize in the production and sale of high-quality breeding animals.

Defined calving season refers to a period, or season, each year during which beef cows are bred to calve. In northern areas of the United States, cows are most often bred to calve during the late winter or early spring to minimize calf loss due to severe winter weather. Some producers prefer fall calving so that calves will be old enough to take advantage of grazing on spring pastures. Year-round calving, or the lack of a calving season, is defined in the National Animal Health Monitoring System as the failure to remove a bull from the herd for at least 30 days (USDA, APHIS, 2010b).

Economies of size is a concept that the average cost of production per unit declines as the number of units produced increases. If so, large operations are able to take advantage of economies of size.

Farm operator refers to the principal operator—the one most responsible for making the day-to-day decisions and running the farm operation.

Farm typology is a farm classification that categorizes farms according to a measure of size, operators' expectations from farming, stage in the life cycle, and dependence on agriculture. The farm typology measure used in this report is:

Rural residence farms are those with annual gross sales below $250,000, including: (1) farms with sales less than $100,000, farm assets less than $150,000, and total household income less than $20,000 per year (limited resource farms); (2) farms whose operators report that they are retired (retirement farms); and (3) farms whose operators report a major occupation off-farm (residential lifestyle farms).

Intermediate farms are those with annual sales below $250,000 and whose operators report farming as their major occupation.

Commercial farms consist of large family farms with annual sales above $250,000 and some nonfamily enterprises organized as cooperatives or nonfamily corporations, and farms with a hired manager.

Feedlot is a type of animal-feeding operation used for finishing cattle prior to slaughter. Cattle are finished, usually on a combination of forage and predominantly grain, to a slaughter weight of 1,000-1,500 pounds. Beef cow-calf producers may feed cattle in a farm feedlot or retain ownership of their cattle being fed off-farm in a commercial feedlot.

Forage quality testing is used to measure such forage components as dry matter, crude protein, and total digestible nutrients. Information from forage quality tests allows cattle producers to better align feed nutrients with animal requirements.

Growth promoting implants are hormones implanted in the ear of beef calves to be absorbed in the blood stream with the purpose of increasing daily gain and feed efficiency.

Information management technologies are systems or processes that provide the information necessary to manage operations more effectively. Examples used on beef cow-calf farms include individual cow records, onfarm computer records, and beef cattle information accessed from the Internet.

Ionophores are antibiotic-like compounds that are fed to beef calves with the purpose of increasing daily grain and feed efficiency.

Net cash farm income is gross cash income less cash expenses. This indicates the cash earnings realized within a calendar year from the sales of farm production and the conversion of assets, both inventories and capital consumption, into cash. Net cash income is a solvency measure representing the funds that are available to farm operators to meet family living expenses and make debt payments (USDA, ERS, 2010c).

Net farm income is net cash farm income less charges for depreciation and the value of noncash benefits for hired labor, plus the value of commodity inventory changes and nonmonetary income. Net farm income is a value of production measure, indicating the farm operators' share of the net value added to the national economy within a calendar year, independent of whether it is received in cash or a noncash form such as increases or decreases in inventories and imputed rental for the farm operator's dwelling (USDA, ERS, 2010c).

Off-farm income is the portion of farm household income derived from off-farm businesses, wages and salaries, interest and dividends, and other sources. Dividends earned by the household are from investments in equities such as stocks or mutual funds. Other sources of off-farm income include pensions, annuities, military, retirement, unemployment, Social Security, veterans' benefits, other public retirement and public assistance programs, and rental income from nonfarm properties (Mishra et al., 2002). Off-farm income can be generated by the farm operator, the spouse, and/or other members of the farm household.

Operating costs of beef cow-calf production are the costs for purchased input items that are consumed during one production period. These are feed; cattle purchased for backgrounding; veterinary services; marketing; custom services and supplies; fuel, lubrication, and electricity; repairs; hired labor; and operating capital.

Ownership costs of beef cow-calf production are the costs associated with the ownership of depreciable assets, such as farm tractors, beef cows, and cattle-handling facilities. These are depreciation, interest, property taxes, and insurance.

Primary occupation is the occupation, farming or otherwise, in which farm operators of beef cow-calf operations spend the majority of their time.

Private pasture/range refers to pasture or range land that is owned or rented by the farm in order to graze beef cattle.

Producer classification is a designation given to farms according to which phases of beef cattle production are conducted on the farm:

Cow-calf only producers sell all calves at or around weaning.

Cow-calf/stocker producers retain calves after weaning and background the calves until they are sold to a feedlot.

Cow-calf/feedlot producers retain calves after weaning and finish all or part of the calves in a feedlot until they are sold as slaughter cattle.

Public grazing land refers to land that is owned by the public sector and managed by Federal, State, or tribal governments. Public grazing land is primarily in large tracts in Western States and is allocated for the use of grazing cattle by ranchers. Fees collected from ranchers for raising cattle on public grazing lands offset costs of range improvements.

Rotational grazing involves dividing the pasture or range into parcels and grazing each in sequence throughout the grazing period. Utilizing rotational grazing can improve livestock distribution while incorporating a rest period for new forage growth.

Stocker calves are young growing animals raised on pasture with very little other feed, with the intention of increasing animal weight and maturity before placing the calves in a feedlot (see backgrounding).

Total economic costs of beef cow-calf production are the full costs of all resources engaged in the beef cow-calf enterprise. These include operating plus ownership costs, opportunity costs for unpaid labor and land, and general farm overhead costs.

ACKNOWLEDGMENTS

Thanks go to Jim MacDonald of USDA's Economic Research Service as well as four other anonymous reviewers for providing indepth reviews of an earlier draft. Thanks also to John Weber and Victor Phillips, Jr. for editorial and design services. Special thanks go to the beef cow-calf producers who participated in USDA's 2008 Agricultural Resource Management Survey and those who collected the data.

REFERENCES

Boggs, D., and E. Hamilton. 1997. *Cow/Calf Analysis: Key Indicators of Profitability*. Proceedings, The Range Beef Cow Symposium XV, Rapid City, SD. December 9-11.

Breiner, S.J., K.M. Boone, D.A. Blasi, S.A. Grau, T.C. Schroeder, B, .B. Barnhardt, R.M. Breiner, and A.M. Bryant. 2007. *Attitudes of Cow-Calf Producers Toward NAIS Are Guarded*. Cattleman's Day Conference Paper. Kansas State University, Manhattan, KS. March.

Fernandez-Cornejo. 2007. *Off-Farm Income, Technology Adoption, and Farm Economic Performance*. Economic Research Report No. 36. U.S. Department of Agriculture, Economic Research Service. January. www. ers.usda.gov/publications/err36/

Gentner, B.J., and J.A. Tanaka. 2002. "Classifying Federal Public Land Grazing Permittees." *Journal of Range Management*. 55:2-11. January.

Golan, E., B. Krissoff, F. Kuchler, L. Calvin, K. Nelson, and G. Price. 2004. *Traceability in the U.S. Food Supply: Economic Theory and Industry Studies*. Agricultural Economic Report No. 830. U.S. Department of Agriculture, Economic Research Service. March. www.ers.usda.gov/ publications/aer830/

Hoder, N., L. Leistritz, D. Nudell, C. Clark, D. Griffith, and T. Jensen. 2007. *Northern Great Plains Beef Production: Production and Marketing Practices of Cow-Calf Producers*. Agribusiness and Applied Economics Report No. 609. Department of Agribusiness and Applied Economics, North Dakota State University. August.

Hoppe, R.A., J.E. Perry, and D. Banker. 2000. *ERS Farm Typology for a Diverse Agricultural Sector*. Agriculture Information Bulletin No. 759. U.S. Department of Agriculture, Economic Research Service. September. www.ers.usda.gov/publications/aib759/

Key, N., and W. McBride. 2007. *The Changing Economics of U.S. Hog Production*. Economic Research Report No. 52. U.S. Department of Agriculture, Economic Research Service, December. www.ers.usda.gov/ publications/err52/

Langemeier, M.R., J.M. McGrann, and J. Parker. 2004. *Beef Cattle Handbook: Economies of Size in Cow-Calf Production*. BCH-8100.

Lawrence, J.D., D. Strohbehn, D. Loy, and R. Clause. 2003. *Lessons Learned from the Canadian Cattle Industry: National Animal Identification and the Mad Cow*. MATRIC Research Paper 03-MRP. Midwest. Agribusiness Trade Research and Information Center, Iowa State University.

MacDonald, J., and W. McBride. 2009. *The Transformation of U.S. Livestock Agriculture: Scale, Efficiency, and Risks*. Economic Information Bulletin No. 43. U.S. Department of Agriculture, Economic Research Service. January. www.ers.usda.gov/publications/eib43/

MacDonald, J.M., E.J. O'Donoghue, W.D. McBride, R.F. Nehring, C.L. Sandretto, and R. Mosheim. 2007. *Profits, Costs, and the Changing Structure of Dairy Farming*. Economic Research Report No. 47. U.S. Department of Agriculture, Economic Research Service. September. www.ers.usda.gov/publications/err47/

McBride, W.D., and N. Key. 2007. *Characteristics and Production Costs of U.S. Hog Farms, 2004*. Economic Information Bulletin No. 32. U.S. Department of Agriculture, Economic Research Service. December. www. ers.usda.gov/publications/eib32/

Mishra, A.K., H.S. El Osta, M.J. Morehart, J.D. Johnson, and J.W. Hopkins. 2002. *Income, Wealth, and the Economic Well-Being of Farm Households*. Agricultural Economic Report No. 812. U.S. Department of Agriculture, Economic Research Service. July. www.ers.usda.gov/ publications/aer812/

Murphy, R.G.L., D.L. Pendell, and G.C. Smith. 2009. "Lessons from the Canadian Cattle Industry for Developing the National Animal Identification System." *International Food and Agribusiness Management Review*, Vol. 12, Issue 2.

Ramsey, R.D. Doye, C. Ward, J. McGrann, L. Falconer, and S. Bevers. 2005. "Factors Affecting Beef Cow-Herd Costs, Production, and Profits." *Journal of Agricultural and Applied Economics* 37, 1(April 2005):91-99.

Schnefp, R. 2010. *Animal Identification and Traceability: Overview and Issues*. Congressional Research Service Report 7-5700. February.

Schulz L.L., and G.T. Tonsor. 2010. "Cow-Calf Producer Preferences for Voluntary Traceability Systems." *Journal of Agricultural Economics* 61: 138-162.

Short, S. 2001. *Characteristics and Production Costs of U.S. Cow-Calf Operations*. Statistical Bulletin No. 974-3. U.S. Department of Agriculture. Economic Research Service. November. www.ers.usda.gov/publications/ sb974-3/

Skaggs, R., and T. Crawford. 2007. "National Animal Identification and the Elephant in the Room." *Rangelands*. April.

U.S. Department of Agriculture, Animal and Plant Health Inspection Service (USDA, APHIS). 2007. *National Animal Identification System (NAIS)—A Users Guide and Additional Resources*. Version 2.0. December.

U.S. Department of Agriculture, Animal and Plant Health Inspection Service (USDA, APHIS). 2008. *Beef 2007-08 Part I: Reference of Beef Cow-Calf Management Practices in the United States, 2007-08*. October.

U.S. Department of Agriculture, Animal and Plant Health Inspection Service (USDA, APHIS). 2009. *Overview Report of the Benefit-Cost Analysis of the National Animal Identification System*. April.

U.S. Department of Agriculture, Animal and Plant Health Inspection Service (USDA, APHIS). 2010a. *Beef 2007-08 Part IV: Reference of Beef Cow-Calf Management Practices in the United States, 2007-08*. February.

U.S. Department of Agriculture, Animal and Plant Health Inspection Service (USDA, APHIS). 2010b. *Beef 2007-08 Part V: Reference of Beef Cow-Calf Management Practices in the United States, 2007-08*. April.

U.S. Department of Agriculture, Economic Research Service (USDA, ERS). 2010a. "Farm Income: Data Files." www.ers.usda.gov/data/farmincome/ finfidmu.htm.

U.S. Department of Agriculture, Economic Research Service (USDA, ERS). 2010b. "Cattle Briefing Room." www.ers.usda.gov/briefing/cattle/background.htm.

U.S. Department of Agriculture, Economic Research Service (USDA, ERS). 2010c. "Farm Income and Costs Briefing Room." www.ers.usda.gov/ briefing/farmincome/glossary/ glossary.htm

U.S. Department of Agriculture, Economic Research Service (USDA, ERS). 2010d. "Commodity Costs and Returns Data." www.ers.usda.gov/data/ costsandreturns/testpick. htm

U.S. Department of Agriculture, National Agricultural Statistics Service (USDA, NASS). 2007. Census of Agriculture. www.agcensus.usda.gov/ publications /2007/full_report/index.asp

End Notes

[1] Includes beef and dairy cattle sales.

[2] All means estimated from Agricultural Resource Management Survey (ARMS) data and presented in this report are weighted by the survey weights.

[3] The 2007 Census of Agriculture measured beef cow inventory on December 31, 2007. The ARMS beef cow inventory used in this comparison reflects the count on January 1, 2008.

[4] Results in this section exclude calves that were retained as replacement breeding stock.

[5] The differences of group means were statistically tested. The differences emphasized throughout this report are statistically significant at a 95-percent level of confidence.

[6] Farm product value discussed in the previous section is the value of all farm products produced in 2008. Gross cash income is the value of products sold in 2008 regardless of when they were produced.

[7] Year-round calving is reported if bulls were not removed from the herd for at least 30 days.

[8] With significant economies of size achieved by operations with 50 or more cows, one would expect operations with 1-19 cows to have much higher costs, especially capital and labor costs, than other operations.

[9] Tags and tagging costs reflect the costs of official identification devices and the application of these devices to the cattle. It was assumed that RFID (radio frequency) eartags would be used for the identification of all cattle.

[10] The survey questions were: Are you familiar with the National Animal Identification System? If yes; Are your premises registered with the NAIS?

[11] Part of this difference is due to NAHMS targeting operations with any beef cows, while farms in ARMS had to have at least 20 cows. NAIS registration is lowest among the smallest operations. NAHMS indicates that only 11.7 percent of beef cow-calf operations with less than 50 cows were registered with NAIS (USDA, APHIS, 2008).

[12] Part of the farm product value was from the production of harvested forages that were fed to cattle. The value of these forages was not counted as part of the value of beef cow-calf production.

In: There's The Beef
Editors: Jonathan L. Ruiz and S. R. Fox

ISBN 978-1-61942-943-7

Chapter 2

CATTLE SECTOR PRODUCTION PRACTICES AND REGIONAL PRICE DIFFERENCES*

Kenneth H. Mathews, Jr. and Rachel J. Johnson

ABSTRACT

Fed cattle from the Southern Plains generally are priced at a premium to Northern and Central Plains cattle. The "North-South price spread" persists because, while production practices in the US cattle sector typically include pasture grazing followed by feeding grain in feedlots, the timing and the duration between each stage may vary. Regional differences include variation with the length and timing of the calving season, the availability and quality of pasture (including the potential for winter wheat grazing), and grain prices. This report describes nuances in regional production practices that underlie the price relationship referred to as "the North-South spread."

Keywords: Backgrounding, calf-fed, cattle feeding, wheat pasture, and yearling

INTRODUCTION

Typically, Southern Plains cattle sell at a premium to northern cattle. While more commonly referred to as the North-South (N-S) price spread, the spread is defined here as the monthly all-grade prices for Southern Plains' (Texas, Oklahoma, and New Mexico) steers minus the prices for Nebraska's steers (for example, USDA, AMS, LM_CT181 and LM_CT183).[1] The N-S spread appears counter-intuitive because higher quality cattle, like those marketed in the Central and Northern Plains, generally sell at a premium to lower quality cattle like those in Southern Plains feedlots. Between 1990 and 2009, monthly Southern Plains fed cattle prices were $0.28 per hundredweight (cwt) higher, on average, than northern cattle prices (fig. 1). Typically, the spread is widest in December-January, is inverted

* This is an edited, reformatted and augmented version of The United States Department of Agriculture publication, LDP-M-202-1, dated April 2011.

in March-April, and narrows during June-August (fig. 2). In addition to seasonal inversions, the N-S spread is sometimes inverted for much longer periods, as it was for much of 1997 and 2003 (fig. 1).

The N-S spread is the result of complex dynamics among a number of differences in cattle sector production systems:

- Timing of production process;
- Quality, age, and size of cattle when marketed;
- Regional marketing conventions; and
- Corn/feed prices in the two regions.
- Regional differences in how these factors vary across production systems and how prices are reported can explain much of the N-S spread.

Cattle Sector Production Systems

Three general groups of production systems for beef cattle characterize the stages from weaning through slaughter with many variations across these systems (fig. 3). The length of time between weaning and feedlot placement distinguishes each production system:

- Calf-fed;
- Intermediate-term (background lot and short-yearling); and
- Long-yearling.

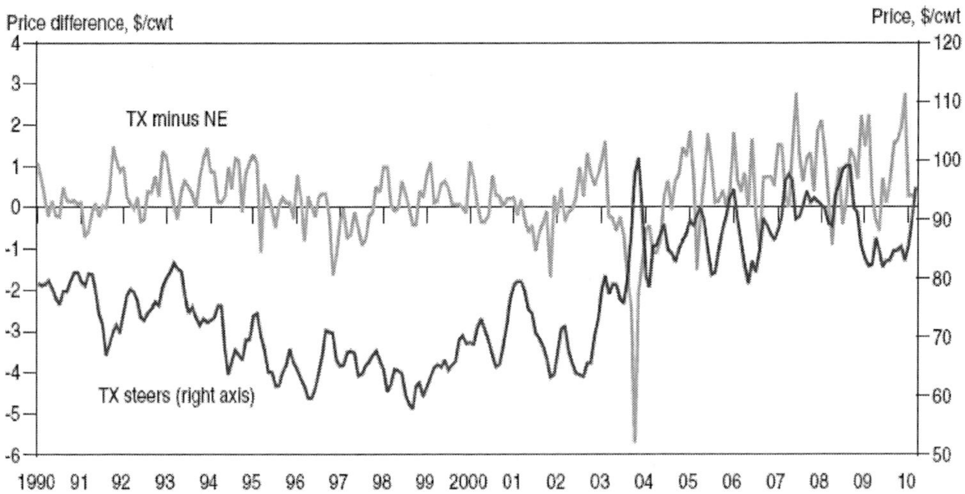

Source: USDA, Economic Research Service calculations based on prices reported to USDA, Agricultural Marketing Service's Mandatory Price Reporting System.

cwt = hundredweight.

TX = Texas prices representative of Southern Plains cattle. NE = Nebraska prices representative of Northern Plains cattle.

Figure 1. Southern cattle often sell at a premium to northern cattle.

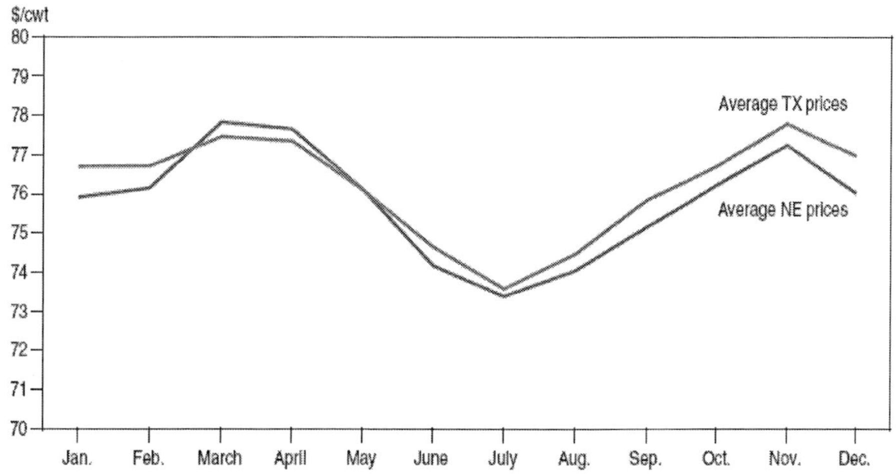

Source: USDA, Economic Research Service calculations based on monthly prices reported to USDA, Agricultural Marketing Service.

cwt = hundredweight.

TX = Texas prices representative of Southern Plains cattle.

NE = Nebraska prices representative of Northern Plains cattle.

Figure 2. Average monthly prices for fed cattle, 1990-2009 $/cwt.

Fed cattle are those placed in feedlots, where they are fed grain and supplements until they reach optimal slaughter weight and grade. Calves are weaned at about 7-8 months of age and can follow one of several alternative interim programs before being placed in feedlots on full (concentrate) feed rations—rations with a high ratio of grain and protein to roughage—until slaughter.

In the most direct system, calves weighing around 600 pounds at weaning can be placed directly on feed (Mark, 2009).

The two primary intermediate-term programs are (1) a backgrounding program[2] that uses supplemental growing rations and (2) a pasture-based short-yearling program that may include supplemental growing rations. In these intermediate-term programs, backgrounded calves and short-yearlings are placed on feed 4-6 months after weaning and will weigh around 700-800 pounds.

In a long-term program, older yearling cattle typically are backgrounded through the winter following fall weaning, and then on pasture the following summer where they grow to 800-1,000 pounds before being placed in feedlots in late summer or early fall (Mark, 2009). This classification—longyearlings—suggests an age dichotomy between feeding younger calves ("calf-feds" and short-yearlings) and older calves ("yearling-feds") and between younger and older calves of the same weight, with discounts often associated with calves purchased to be placed on feed over yearlings of the same weights (discussed later).

At any point in time, cattle on feed come from all production systems. Feeding periods can range from less than 90 days to more than 300 days. Generally, the older the cattle are at placement, the shorter the feeding period because they are larger and the feed they consume is used primarily to achieve optimal "finish." Less efficient cattle—those consuming relatively more feed per pound gained or those gaining at slower rates—can be on feed for longer periods, regardless of age.

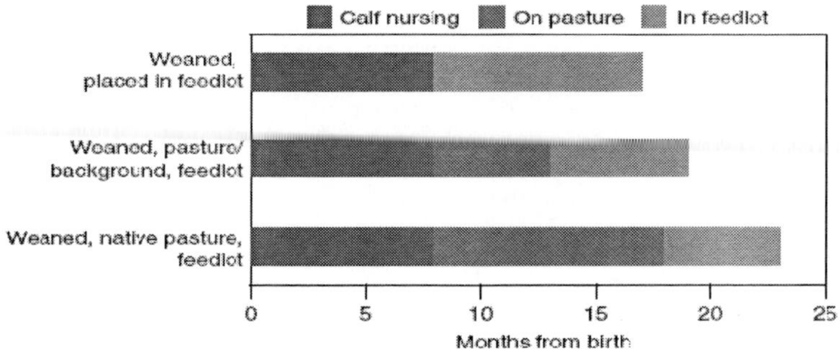

Source: USDA, Economic Research Service.

Figure 3. Timeline for beef cattle production systems.

Calf-Fed System

Most US calves are born in the spring and are weaned in October-November. Warmer climates in the South offer greater access to adequate cool-season pastures, providing quality pasture for milk production to support calving throughout the year. Calf-fed cattle are generally acclimated to feeding regimes with less concentrated rations for 2-3 weeks before going on full rations. If calves are on a less concentrated, or growing, ration for several months, they are considered part of the backgrounding lot system—a variation of the short-yearling system (discussed later)—because of the calf's age and weight when placed on feed. Calf-feds typically are smaller, have smaller digestive systems, and are also still growing. Because of these growth characteristics, calves placed on feed at or shortly after weaning will often be on feed longer than the average 5- to 6-month feeding period (typically 8-9 months). Calves placed shortly after weaning in October-November will be marketed as fed cattle from May to early July (Meyer and Steiner, 2007; Mark, 2009), when prices begin seasonal declines, and will weigh less than most cattle in yearling systems.

The number of cattle on feed for more than 120 days typically peaks during the May-July marketing period. This benchmark indicator denotes how efficiently feedlot managers are marketing cattle before they become too fat, or "over-finished," and as they become market ready. Some latitude is given, however, to interpreting this commonly discussed benchmark because calffeds are typically on feed for more than 120 days.

Again, because of their smaller size and growth patterns, calf-feds often perform better in feedlots in terms of growth and feed conversion than older calves and yearlings. When corn prices are low, managers may place more calves on feed at weaning because calf-feds gain weight more efficiently on corn-based diets (Griffin et al., 2008). Corn prices in Nebraska are consistently lower than corn prices in the Southern Plains partly because of transportation costs. As a result, managers may feed more calves in the northern region. Due to their smaller carcasses and their tendency to become excessively fat if kept on feed too long (Griffin et al., 2007), calf-feds are often discounted more than yearlings of the same finish. Calf-feds often produce meat that is more tender (Brewer et al., 2003), but they may also yield less meat as a percentage-of-live-weight basis than yearlings, making them slightly more expensive to process (Smith, 2005).

Of the three production systems discussed in this report, the long-yearling system keeps cattle on pasture for the longest time before placing them in feedlots.

Intermediate-Term Programs

Short-yearlings—backgrounded in dry lots (pens generally devoid of or with limited pasture) or on pasture prior to placement in feedlots—are placed on feed at or near 1 year of age, where they remain for 5-6 months. Prior to entering the feedlot, they will have eaten rations (if backgrounded) or grazed pasture of sufficient quality for them to gain 1-2 pounds per day for 4-6 months. In the wheat-growing areas of the Southern Plains, cattle commonly graze on wheat or other small grain pastures throughout the winter and gain, on average, 2 or more pounds per day. In the Central and Northern Plains, cattle grazing on either native pasture, corn stalks or other stover, or harvested forages must be fed energy/protein supplements to gain adequate weight during the winter.

Long-Yearlings

Long-yearlings may be slower growing calves that simply did not reached sufficient weight for placement in feedlots, or they may have spent the winter on a low-quality pasture that prevented sufficient growth. Many northern pastures are at their nutritional peak during spring and early summer. Calves on these pastures will ordinarily gain enough weight to enter feedlots from late summer to early fall. Long-yearlings can weigh over 1,000 pounds when placed, but usually weigh closer to 800-900 pounds.

Because they are larger and can use feed more efficiently for "finish" rather than growth, long-yearlings are generally able to reach market-ready condition sooner. Long-yearlings, however, also tend to continue growing and can become overweight at slaughter (Griffin et al., 2007), sometimes resulting in discounts. These cattle may be marketed after being on feed less than 5 months, weighing 1,400 pounds or more. Long-yearlings are capable of rapid weight gain early in the feeding period—compensatory gains at as much as twice the "normal" rate of growth—because of "pent up" growth potential that was not completely fulfilled on growing regimes.

In one Nebraska study, a calf crop was divided into two groups at weaning. The researchers compared calf-fed and long-yearling systems. Calf-feds—at the heavy end of the calf crop—entered feedlots weighing an average of 642 pounds, were on feed an average of 168 days, and were marketed at an average weight of 1,282 pounds (Griffin et al., 2007). Calves at the light end of the calf crop weighed 526 pounds and were placed on pasture from fall to the following September. They entered feedlots as long yearlings weighing an average of 957 pounds and were on feed for only 90 days before being marketed at an average weight of 1,365 pounds.

In another aspect of the same analysis, "[p]rofitability of yearlings versus calf-feds increased $4-6 [per head] for each $1/bushel increase in corn price" (Griffin et al., 2008). While there were no differences in marbling scores or yield grades, yearlings gained more weight on less feed in the feedlot than did calf-feds, but required more total time from weaning to slaughter (Griffin et al., 2007).

SOUTHERN PLAINS CATTLE SELL AT A PREMIUM TO NORTHERN PLAINS CATTLE

Northern cattle grade higher, on average, than southern cattle, and higher quality cattle are priced at a premium to lower quality cattle. As such, the observed N-S spread runs counter to the typical premiums paid for quality. For example, in 2009, the N-S spread between Nebraska (northern) prices and Texas (southern) prices averaged $1.12 per cwt across all grades and qualities. In 2009, 46 percent of negotiated purchases of Nebraska fed cattle were in the 65- to 80-percent Choice range, whereas 92 percent of TexasOklahoma-New Mexico cattle were in the lower quality 35- to 65-percent Choice range (USDA, AMS, LM_CT175 and LM_CT173).[3] Two factors partially explain why the southern premium might exist despite the average difference in quality: The distance between feedlots and packers and the difference in how shrinkage[4] is discounted.

Although information has been added to the markets for fed cattle, not all the information being reported to the Mandatory Price Reporting System is reported in ways that are necessarily directly comparable. Reported prices do not reflect differences in marketing practices and channels; nor are differences in transportation arrangements reported. For example, packers may pay transportation costs in some cases, cattle owners may pay transportation costs in other cases, or the costs may be split. The distance between the feedlot and the packer is also a factor.

At times, especially when fed cattle are relatively plentiful, cattle feeders and packers may not agree on prices, so the cattle remain on feed longer and become "over-finished." These cattle then shift into higher yield grades[5] (grades 4 and 5) and weigh more. At the extreme, these heavier cattle often sell at discounted prices.

On average, northern cattle are generally 100 pounds or more heavier when marketed than southern cattle because southern cattle often come from sources with lighter average placement weights (Southeastern United States and Mexico). In 2009, Nebraska fed cattle averaged 1,398 pounds (live weight across all quality categories) versus 1,286 pounds for Texas-Oklahoma-New Mexico cattle.

Other subtle differences in information are reported to the AMS Mandatory Price Reporting System, where the data are aggregated as cattle are sold.

For example, fed cattle are sold out of feedlots in a number of ways. Some cattle are sold on a dressed basis as hanging carcasses, usually with no adjustment for shrinkage. Other cattle are sold as live animals, usually with an adjustment for shrinkage.

In the eastern portions of Northern and Central Plains cattle feeding areas, cattle are often sold with a 3-percent shrink discount— packers pay based on actual live weight adjusted down by 3 percent to account for weight lost in transit from the feedlot to the packer. Southern Plains cattle and some cattle from western portions of Northern and Central Plains cattle feeding areas are often sold with a 4-percent shrink discount. If all other factors remain equal, such as live weight, quality, and distance to the packer, northern cattle would sell at a slightly lower price per cwt than southern cattle simply due to varying shrink factors (table 1).

Table 1. The hypothetical effect of shrink on prices, given constant live weight and quality

	Live weight	Shrunk weight	Price per head	Price per cwt
Hypothetical 3-percent shrink	------Pounds-----		---------Dollars-------	
(e.g., NE)	1,300	1,261	1,170	92.78351
Hypothetical 4-percent shrink				
(e.g., TX)	1,300	1,248	1,170	93.75
Difference				0.966495

Source: USDA, Economic Research Service.

Dynamics of the North-South Price Spread and Cattle Production Systems

The very nature of cattle production systems play a role in the N-S price spread:

1) The relative supply of feeder cattle and fed cattle (by season and region);
2) The age, size, and price of cattle placed in feedlots (by season);
3) The occurrence of dry periods (by region); and
4) The availability of various feedstuffs like corn gluten feed or distillers' grains (by region).

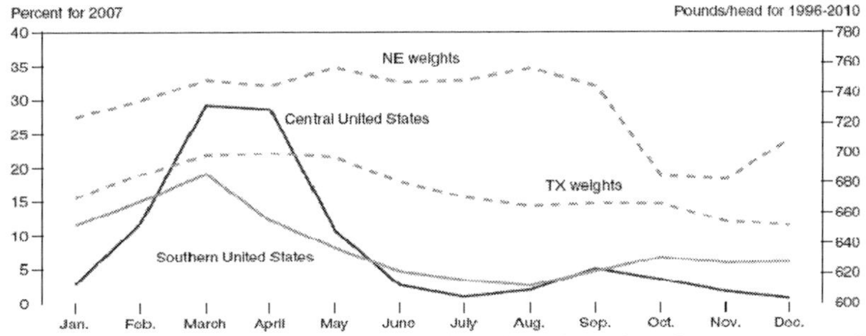

Source: USDA, Economic Research Service calculations based on data from USDA, Animal and Plant Health Inspection Service, Veterinary Services, Beef 2007-08, Part II: Reference of Beef Cow-calf Management Practices in the United States, 2007-08; National Agricultural Statistics Service, Cattle On Feed reports, various issues.

TX = Southern Plains cattle.

NE = Northern Plains cattle.

Figure 4. Monthly calving distribution and average placement weights.

Supplies of fed cattle vary between northern and southern feedlots, altering the price spread. Regional calving patterns, grazing options, and feedlot placement weights underlie regional differences. Nationwide, 54 percent of the yearly calf crop is born during March-May (USDA, APHIS, 2009). The spring calving tendency holds more rigidly in the North—with its harsher winters and shorter summer pasture-grazing seasons—where 69 percent of calves are born during March-May (fig. 4).

Some of these spring calves are placed in feedlots in October-November (figs. 5 and 6). Milder winters allow a for lengthier spring calving season in the South and a greater number of calves to be born in the fall.

Thus, weaning and placement at weaning in the South can be spread over a longer period. As a result, marketing of southern calf-feds is distributed more throughout the year than in the North.

A greater number of grazing options are available for feeder cattle not placed on feed immediately after weaning in the South through the winter months because of milder winter temperatures and the availability of cool-season grasses. More feed options offer southern producers greater flexibility in calving seasons and grazing programs. In the South, cattle can often graze winter wheat pasture, generally from October through mid-March. Feedlot placements increase sharply in March when cattle have to be removed from wheat pastures to prevent inroads into grain production (see fig. 6). Another peak occurs in May when wheat pastures not intended for harvest have been "grazed out" and cattle are removed from these or other cool-season pastures (or other backgrounding programs) and placed in feedlots. Northern cattle placement in feedlots increases again in May, mostly from short-yearling backgrounding programs, possibly as a means of filling pen space vacated by earlier marketings of calf-feds.

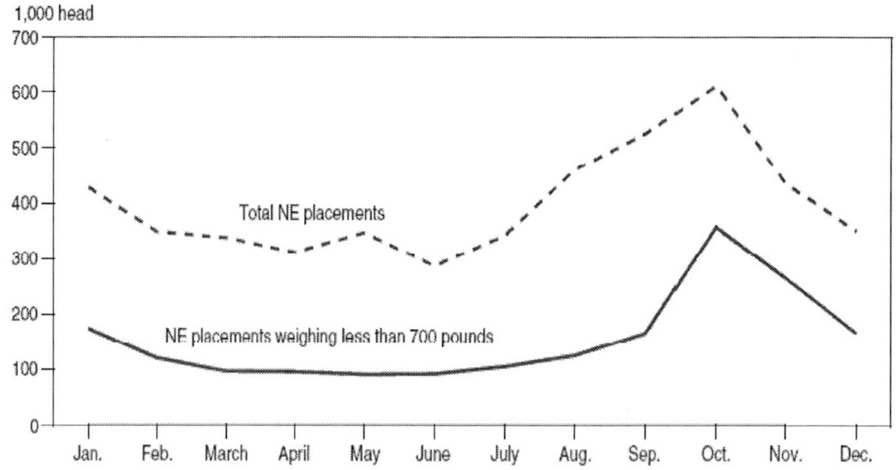

Source: USDA, Economic Research Service calculations based on data from USDA, National Agricultural Statistics Service, Cattle On Feed reports, various issues.

Figure 5. Total Nebraska placements and those under 700 pounds, averaged over 1996-2010.

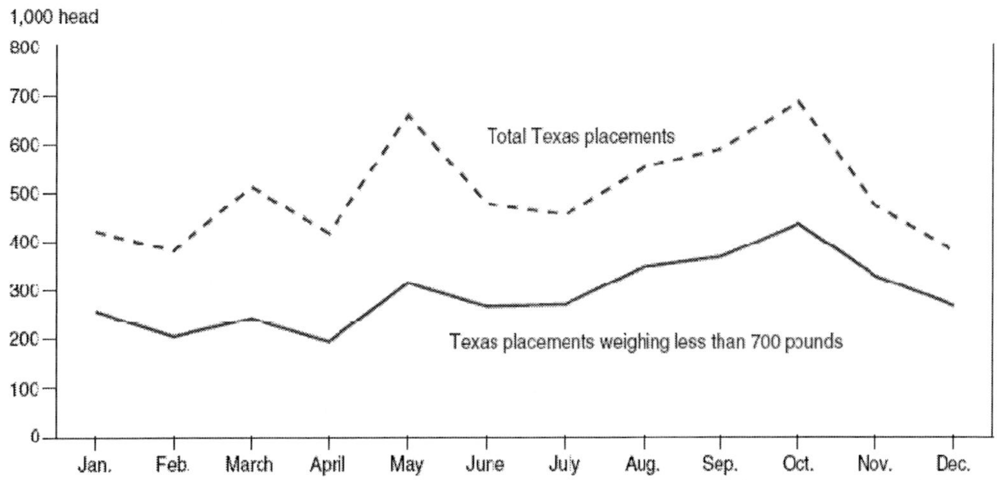

Source: USDA, Economic Research Service calculations based on data from USDA, National Agricultural Statistics Service, Cattle On Feed reports, various issues.

Figure 6. Total Texas placements and those under 700 pounds, averaged over 1996-2010.

On average, the peak in northern feeder cattle placement occurs in October (see fig. 5). A relatively large proportion of the cattle entering feedlots in the North from October to November are calf-feds. These cattle generally weigh less than 700 pounds and will be marketed primarily during May-July.

In a typical year, however, more 700-pound or heavier cattle (long-yearlings) are placed in the North than in the South.

Placement of these long-yearlings peaks in September-October, and marketing of these cattle from feedlots begins in late February (see figs. 5 and 6). Conversely, more cattle weighing less than 700 pounds are typically placed in TX than NE, with peak placement in April-June. These placements may be calf-feds, calves carried through the winter, or calves that have grazed out wheat pastures.

Placements in this region and weight category also are likely to include a number of calves imported to the United States from Mexico.

Heavier calves in the South placed in February-March likely are removed from wheat pasture early enough to prevent damage to the grain crop. Heavier cattle placed in April-June are generally short-yearlings that have grazed out wheat pasture or another small grain crop. These April-June placements are marketed as fed cattle beginning in late summer. Long-yearlings in both the North and South are placed in September-October and sold from feedlots beginning in late February (figs. 5, 6, and 7).

As a result of these placements, supplies of market-ready fed cattle increase earlier in northern feedlots in March-May. Fed cattle marketings in the North are elevated in May-October. Fed cattle prices generally reach a seasonal peak in March or April, on average, because fewer market-ready cattle are available during the first quarter (see fig. 7).

This peak occurs just after long-yearlings in the North and South have been marketed and calf-fed placements have not yet reached market finish. Prices peak again in November, with reduced supplies of fed cattle resulting from low feeder cattle placements during spring and summer.

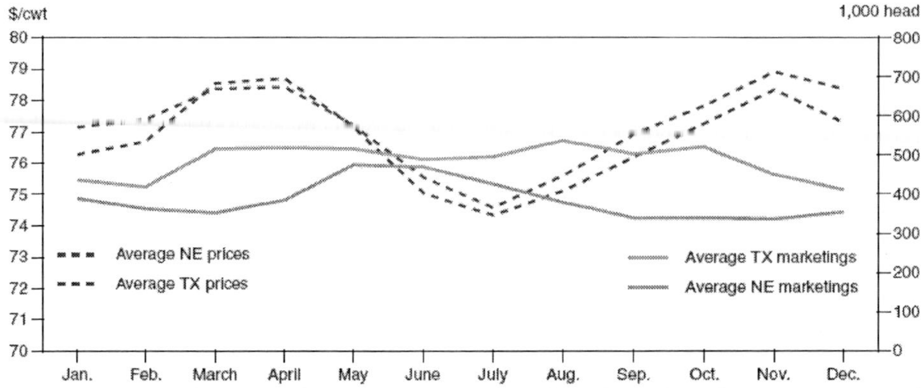

Source: USDA, Economic Research Service calculations based on data from USDA's Agricultural Marketing Service reports on monthly prices.

cwt = hundredweight.

Figure 7. Monthly prices and marketings for fed cattle from Texas and Nebraska feedlots of 1,000 head or more, 1996-2010 average.

Seasonally low prices occur in June-August with another, less pronounced seasonal dip in December-January. The June-August low in prices is due to the relatively large supply of fed cattle available during the summer, including calf-feds from the North and South and early finishing yearlings in the South that were removed from wheat pasture and cool-season forage.

Price Spread Inversions Occur Seasonally

The average monthly price differential between Texas' and Nebraska's (TX-NE) cattle decreases sharply in the early spring, becoming negative in March-April when larger numbers of fall-placed calf-feds and long-yearlings come to market as fed cattle (fig. 8). Relative scarcity for long-yearlings in NE results in an average price premium for NE cattle compared with TX cattle prices during March-April, making the TX-NE spread negative (figs. 7 and 8). During these months, a similar number of calf-feds come from both North and South, but fewer preferred yearlings are coming to market in the North. Packers in the North who want to process more yearlings than are locally available and who wish to maintain slaughter levels have three options:

1) They must bid more for northern yearlings due to their relative scarcity;
2) They must pay more for northern calf-feds; or
3) They must pay transportation costs for southern cattle.

The TX-NE spread is greatest in December-February, when fewer summer-placed cattle are marketed (figs. 7 and 8). The observed reduction in marketings during the first quarter (January-February in TX and January-March in NE) are due to low placement levels from the previous summer.

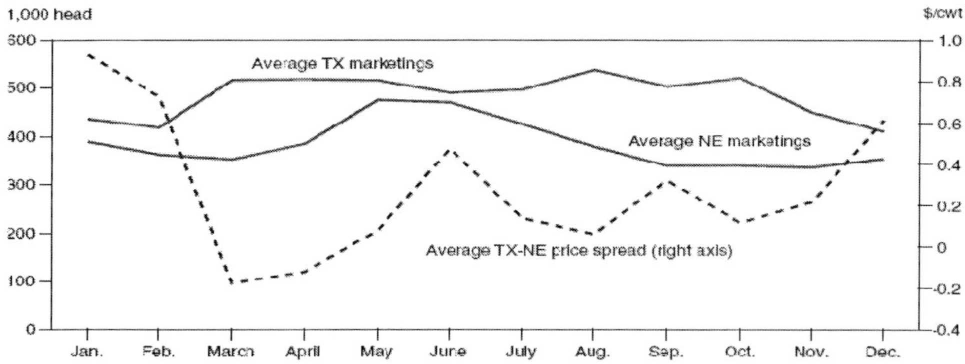

Source: USDA, Economic Research Service calculations based on data from USDA's Agricultural Marketing Service reports on monthly prices; placement data from USDA, National Agricultural Statistics Service, Cattle On Feed reports, various issues.

cwt = hundredweight.

NE = Nebraska cattle.

TX = Texas cattle.

Figure 8. Marketings from feedlots of 1,000 head or more, December 1995-December 2010 average.

During December-January, Southern Plains feedlots have fewer heavier weight (yearling) cattle relative to northern feedlots, prompting southern packers to bid more for these cattle. This supply adjustment occurs just before the bulk of calf-feds enter the market (May-July).

OTHER HISTORICAL PRICE-SPREAD INVERSIONS

Inversions in the price spread between NE and TX cattle do occur outside the seasonal inversion in March-April. For example, in 2003, imports of Canadian feeder and slaughter cattle were halted between May 2003 and July 2005 due to a Canadian case of bovine spongiform encephalopathy (BSE). This event occurred as US cattle inventories were well into cyclical liquidation and were already low. Since many Canadian feeder cattle are typically fed in northern US feedlots,[6] feeder cattle placements in northern feedlots for 2003 were below normal.

At the same time, Mexican cattle imports were increasing, with a large proportion being fed in and marketed from southern feedlots, increased regional supplies. This supply flux resulted in an atypical shift in already reduced supplies of fed cattle available for slaughter from northern feedlots and increased prices relative to southern fed cattle prices. This shift reversed the TX-NE spread (a premium for NE over TX cattle) for all but 2 months (June-July 2004) from March 2003 to August 2004.

The price spread was also inverted for 13 months from October 1996 through October 1997, although the reason is not as clear. Placements of heavier feeder cattle were higher than usual in the Southern Plains compared with Northern and Central Plains placements, which could have given a price edge to northern markets. Also, in fall 1996, feeder cattle imports from Mexico were below those of the previous year, while slaughter steer and heifer imports from Canada were relatively high year to year, which could have reduced the availability of heavier finished cattle in TX and given heavier finished cattle in NE a price advantage.

Finally, cattle-on-feed inventories rose as seasonal grain prices declined from record high levels set in summer 1996. Marketings at or below those of the previous year supported fed cattle prices, and cattle placed on feed in late fall 1996 and winter 1996 could be hedged at profits approaching $100 per head.

Conclusion

Production practices in the cattle sector reflect variations in the timing and duration of calving and the use of pasture before placement in feedlots. Some of these differences contribute to regional price differentials. For example, fed cattle prices in the Southern Plains are typically at a premium to Northern and Central Plains fed cattle. A price difference persists, in part, because of discounts associated with the heavier average market weights of northern cattle and despite the premiums associated with the higher average quality of northern cattle compared with marketings in the Southern Plains. Seasonal factors and aberrant market situations produce premium price inversions for southern fed cattle.

Acknowledgments

Thanks go to Darrell Mark (University of Nebraska), Jim Robb (Livestock Market Information Center), James Hansen (Economic Research Service), Greg Pompelli (Economic Research Service), Paul Wescott (Economic Research Service), Sherry Wise (USDA, Agricultural Marketing Service), and Shayle Shagam (USDA, Office of the Chief Economist) for their helpful review comments and insights. Thanks also go to Angela Anderson and Cynthia A. Ray for their editorial and layout expertise.

References

Brewer, P., C. Calkins, R. Anderson, T.J. Klopfenstein, and R. Rasby.

"Carcass and Palatability Characteristics of Calf-fed and Yearling Finished Steers," *2003 Nebraska Beef Cattle Reports*, University of Nebraska-Lincoln, Animal Science Department, pp. 68-70, 2003.

CanFax Weekly Summary, *Cattle Market Outlook and Analysis* Vol. 38(9), February 29, 2008.

Griffin, W.A., T.J. Klopfenstein, G.E. Erickson, D.M. Feuz, and J.C. MacDonald. "Comparison of a Long Yearling System and Calf-fed Performance and Economics," *2007 Nebraska Beef Cattle Reports*, Extension Publication MP-90, University of Nebraska-Lincoln, Animal Science Department, 2007.

Griffin, W.A., T.J. Klopfenstein, G.E. Erickson, and D.R. Mark. "Evaluation of Calf-fed and Long Yearling Production with Increasing Corn Price," *2008 Nebraska Beef Cattle Reports*, Extension Publication MP-91, University of Nebraska-Lincoln, Animal Science Department, 2008.

Mark, D.R. *In the Cattle Markets*, University of Nebraska-Lincoln, Department of Agricultural Economics, September 28, 2009.

Mark, D.R. Personal communication, December 22, 2009.

Meyer, S., and L. Steiner. *Daily Livestock Report*, Chicago Mercantile Exchange, July 23, 2007.

Smith, T. "Calf-fed Trend: Will the Trend Toward More Calf-feds Continue? How Might it Affect Producers?" *Beef Quality Connection*, May 3, 2005.

US Department of Agriculture, Agricultural Marketing Service. *Nebraska Yearly Weighted Average Cattle Report - Negotiated Purchases, Recap for 2009*, LM_CT175, January 7, 2010a.

US Department of Agriculture, Agricultural Marketing Service. *Texas/ Oklahoma Yearly Weighted Average Cattle Report - Negotiated Purchases*, Recap for 2009, LM_CT173, January 7, 2010b.

US Department of Agriculture, Agricultural Marketing Service. Nebraska *Monthly Weighted Average Cattle Report - Negotiated Purchases*, LM_ CT183, various issues.

US Department of Agriculture, Agricultural Marketing Service. *Texas/Oklahoma Monthly Weighted Average Cattle Report - Negotiated Purchases*, LM_CT181, various issues.

US Department of Agriculture, Animal and Plant Health Inspection Service, Veterinary Services. *Beef 2007-08, Part II: Reference of Beef Cow-calf Management Practices in the United States, 2007-08*, February 2009.

US Department of Agriculture, National Agricultural Statistics Service. *Cattle On Feed* reports, various issues.

End Notes

[1] In what follows, the use of "NE" for Northern and Central price-related designations and "TX" for Southern Plains designations is maintained as a shorthand reference to USDA's Agricultural Marketing Service (AMS) designations for the separate price series that typically define the N-S spread.

[2] Backgrounding prepares young cattle for a feedlot, allowing them to grow and getting them accustomed to confinement facilities and feeds.

[3] USDA's Agricultural Marketing Service is the agency responsible for grading meat. Grading for quality considers the tenderness, juiciness, and flavor of meat at eight levels: Prime, Choice, Select, Standard, Commercial, Utility, Cutter, and Canner (in order from highest to lowest quality).

[4] Shrink results from the stress cattle experience during processing, transporting, and marketing.

[5] The yield grade of a beef carcass is determined by a formula that takes into account the amount of external fat over the ribeye muscle; the amount of kidney, pelvic, and heart fat; the area of ribeye muscle; and the carcass weight. Yield grades range from 1 through 5, with 1 having the least fat and 5 being the fattiest. The lower yield grades (1 and 2) generally have higher dressing percentages (more meat per carcass as a ratio of carcass weight to live weight), while the higher yield grades (4 and 5) have lower dressing percentages. Cattle with higher dressing percentages usually receive higher prices.

[6] CanFax reports that, in 2007, 81 percent of feeder cattle exported to the United States were fed in Nebraska (39 percent), Colorado (21 percent), Iowa (11 percent), and Washington (10 percent), compared with 74 percent fed in those same States in 2006 (CanFax Weekly Summary, Cattle Market Outlook and Analysis Vol. 38(9), February 29, 2008).

In: There's The Beef
Editors: Jonathan L. Ruiz and S. R. Fox

ISBN 978-1-61942-943-7

Chapter 3

WHERE'S THE (NOT) MEAT? BYPRODUCTS FROM BEEF AND PORK PRODUCTION*

Daniel L. Marti, Rachel J. Johnson and Kenneth H. Mathews, Jr.

ABSTRACT

Animal byproducts contribute to the bottom line of the U.S. meat industry. Byproducts (edible offal (including variety meats), inedible offal, hides and skins, blood, fats, and tallow) include all parts of a live animal that are not part of the dressed carcass and constitute about 30 percent of the liveweight of hogs and about 44 percent of the live-weight of cattle.

Byproducts from animal slaughter provide raw materials used in pharmaceutical, cosmetic, household, and industrial products.

Exports of edible offal also contribute to the value and profitability of the U.S. meat processing industry in a way that leads to higher prices for livestock producers, as byproducts account for more than 23 and 35 percent of the volume of beef/veal and pork exports, respectively.

Regression analysis indicates that a 10-percent increase in the steer byproduct drop value adds a 1-percent increase in the five-area weighted average price for all grades of steers.

U.S. exports of beef/veal and pork edible offal have increased in recent years, mostly due to population growth, income growth, and consumer preferences for variety meats, especially in Asia. Income growth in the global marketplace, however, may have varied effects on the consumption and trade of variety meats.

Keywords: Animal byproducts, offal, edible, inedible, variety meats, hogs, cattle, trade

* This is an edited, reformatted and augmented version of the United States Department of Agriculture publication, LDP-M-209-01, dated November 2011.

INTRODUCTION

Animal byproducts, or offal, may not make up the real "heart" of the U.S. meat industry, given that the multiple uses for inedible animal byproducts are not often considered. And, variety meat dishes, such as beef tongue or pig heart, do not typically grace Americans' dinner tables.

Nonetheless, offal derived from beef and pork slaughter contributes to the bottom line of the U.S. meat industry. During 2001-10, edible animal byproducts (byproducts that do not include hides and other inedibles) averaged 23-35 percent of the volume of U.S. pork and beef/veal exports and 14-19 percent of the value.

Animal byproducts include all parts of a live animal that are not part of the dressed carcass. Produced jointly in the process of harvesting meat from the animal, byproducts constitute an estimated 30 percent of the liveweight of hogs and about 44 percent of the liveweight of cattle.

In the United States, animal byproducts fall into three categories: hides, inedible offal, and edible offal, with variety meats being a subcategory of edible offal. Based on live value, byproducts account for more than 10 percent of the value for cattle and more than 6 percent of the value for hogs. Tallow accounts for about 20 percent of the value of live cattle, and lard accounts for about 9-17 percent of the value of a live hog. Although hides/skins and rendered products contribute to the value of cattle and hogs, this report focuses primarily on offal products.

Hides account for 30-75 percent of the byproduct drop value for cattle but very little of the drop value for hogs.[1]

The use of animal byproducts dates back to early civilization, with hides used for clothing and intestines used for food containers. The first recorded use of a cleaning compound, a soap made from animal fat and lye, was in the first century A.D. Early U.S. history also records production and uses for hides and tallow in the Western United States (Ockerman and Hansen, 2000).

Animal byproducts provide many of the raw materials used to make pharmaceutical, cosmetic, household, and industrial products. These products, along with hides and well-known variety meats, such as liver, tongue, hearts, and feet, all add value to the U.S. meat industry. But just how important are they and what role do they play in export trade?

WHAT ARE ANIMAL BYPRODUCTS?

U.S. consumers have come to depend on the wide array of products made from animal byproducts (see app. A). Among countries, the product mixes comprising the edible and inedible offal categories can vary slightly but are mostly consistent. Inedible animal byproducts—those typically considered or mandated as not edible—are the primary raw materials used in the manufacture of a broad assortment of industrial, household, cosmetic, pharmaceutical, and medical supplies, in addition to such products as lubricants, plastics, soaps, glycerin, and gelatins (Aberle et al., 2001). In addition to having domestic value, edible offal contributes as much as one-fourth of the volume of U.S. beef exports and one-fifth of the volume of U.S. pork exports (Marti and Johnson, 2010).

Inedible Byproducts

Inedible animal byproducts include hide or skin, hair, horns, teeth, fats, bone, ligaments and cartilage, feet, glands, blood, and lungs. Although the share varies among individual animals, cattle hides account for about 75 percent of the byproduct value of a beef animal, much more than pork skins contribute to the value of a hog. U.S.-produced hides are often exported to China, Hong Kong, South Korea, Taiwan, Mexico, and the European Union, where they are tanned and processed into leather for shoes, purses, clothing, car seats, and other items.

Some inedible offal, along with normally edible offal that has been deemed unsuitable for human consumption, bones from meat processing, and cattle that are unsuitable for human consumption (nonambulatory and other condemned cattle), is rendered for use in the industrial, cosmetic, and feed manufacturing industries. Processors render products by heating animal tissues and skimming off the fats and oils. Both the skimmed fats and oils and the residual materials are converted to materials that have economic value, generally as inputs into further manufacturing processes. Rendering also helps minimize the release of animal tissues into the environment as potential biological hazards (Aberle et al., 2001; Danilevici et al., 2009). According to Prokop (1996), about 40 percent of the weight of a live beef animal goes through the rendering process. Rendered meat and bone meal is a valuable source of protein in pig and poultry feed, pet foods, and even fertilizer. The use of meat and bone meal made from cattle and sheep (and other ruminants), however, is restricted by regulations designed to reduce the risk of bovine spongiform encephalitis (BSE) (see Mathews, 2008; Mathews et al., 2006).

Animal byproducts also are important to the development and ultimate availability of modern human medicines. Glands removed from livestock at slaughter, such as the adrenal, parathyroid, pituitary, thyroid, and thymus glands; ovaries; pancreas; and testes provide many of the hormones and enzymes used in the medical field (Aberle et al., 2001). Among the medicines that can be obtained from animal glands are epinephrine, estrogens, progesterone, insulin, trypsin, parathyroid hormone, adrenocorticotropic hormone (ACTH), somatotropin, thyroid stimulating hormone, testosterone, thyroxin, and thymosin. Serums, vaccines, antigens, and antitoxins are also derived from many food-animal tissues acquired both during slaughter and processing of the animal (Pearl, 2005). Purified animal blood is fractionated into numerous products, including thrombin, used for blood coagulation agents and skin-graft procedures; fibrin, used in surgical repair of internal organs; and fibrinolysin, used to help heal minor burns or as a wound-cleaning agent.

Certain parts from pigs and cattle are used for xenotransplantation, the insertion of tissue from one species into another. Skin, brain cells, insulin, heart valves, and lungs from pigs are all used for human transplants as well (Goodlight, 2010). Pig skins are used for initial treatment for burn patients, while bone cartilage and bone fragments are used as substitutes for diseased or damaged human tissue parts (Pearl, 2005). Intestines provide surgical ligatures; blood provides albumen, amino acids, fetal serum, and thrombin; bones provide calcium and phosphorus; and other inedible offal provides liver extracts, bile extract, cortisone, heparin, cholesterol, rennet, and pepsin. In many of these treatment uses, no other synthetic products function or perform equally well. In other cases, the extracts of animal byproducts have provided the scientific basis for the development of synthetic substitutes, such as insulin and other pharmaceutical products (Pearl, 2005).

Animal byproducts are also a major contributor to the growth and expansion of the pet food industry (Corbin, 1992). Animal byproducts in pet food diets are good sources of digestible protein, fat, vitamins, and minerals and, historically, have supplied the majority of these nutrients for pets. For example, animal byproducts account for 25-40 percent of the dry matter in premium dog diets (Murray et al., 1997).

Edible Byproducts

The edible byproducts from slaughtered animals are segregated, chilled, and processed. These products include livers, hearts, tongues, tails, kidneys, brains, sweetbreads (the thymus and/or pancreas gland, depending on an animal's age), tripe (stomach), melt (spleen), chitterlings and natural casings (intestines), fries (testicles), rinds, head meat, lips, fats and other trimmings, blood, and certain bones. Typically, edible byproduct yield is around 12 percent of liveweight from cattle and about 14 percent of liveweight from hogs when pork rinds are included (Ockerman and Hansen, 2000, pg. 23).

Edible byproducts can be categorized into variety meats or edible fats and oils. Edible organs and glands—brains, hearts, kidneys, livers, melts, sweetbreads, tongues, and chitterlings—along with oxtails are categorized as variety meats and are usually sold with minimal processing. Intestines and cheek meats are usually processed further, often into sausages and other processed meat products. Extra trimmings and tails are used in soups and bouillons; extra trimmings, blood, stomachs, and intestines can be sausage ingredients or casings. Certain stomach parts also provide rennet for cheese making. Gelatin is used in such products as ice cream and jellied foods and is produced in part from bones and skins (Aberle, 2001).

While many of the variety meats are typically fatty and higher in cholesterol, they are also good sources of essential vitamins and nutrients:

- Livers are high in vitamin A, iron, zinc, B vitamins, vitamins C and D, copper, and fatty acids.
- Hearts contain large amounts of iron and are a good source of selenium, zinc, phosphorous, niacin, and riboflavin, but they are very low in sodium.
- Brains are rich in niacin, phosphorus, B_{12}, and vitamin C.
- Tripe contains abundant protein and B12.
- Sweetbreads are very high in vitamin C.
- Kidneys are high in protein and contain riboflavin and niacin.
- Tongues are a good source of B_{12} but are low in sodium.
- Hog feet are low in sodium but are a good source for selenium.

In some countries, variety meats are considered delicacies and are the basis for many traditional dishes; in other countries, their consumption is associated with low-income populations (Halstead, 1999). Demand for variety meats is especially strong in many Asian nations. In China, many recipes call for sharp-tasting variety meats rather than muscle cuts, which are considered bland (Hayes, 1997); cow tongues are considered expensive delicacies in Japan; and sliced beef feet are used for soup in South Korea. Tongue and liver are used in many Mexican dishes, such as putzaze (tripe and liver with tomatoes), lengua (tongue with green chilies), and menudo norteña (tripe soup).

In Russia and Egypt, two of the world's leading importers of edible offal (head meat, liver, heart, kidney, and tongue), variety meats are more commonly consumed by lower income households and are used as an inexpensive way to obtain high-quality protein and nutrition (Kamenski, 2006).

Byproduct Production

Estimates of total U.S. byproduct production by species are not publicly available, although limited data are available from the U.S. Department of Commerce and *Render Magazine* (National Renderers Association) for some fats and oils and other rendered products from all species.

USDA's Agricultural Marketing Service (AMS) also reports data in its *Pork and Beef Variety Meats Report* on the number of 40,000-pound loads of various byproducts that are sold each month as well as the average value and weight of many of these specific items. The items detailed in the AMS report include only those sold directly for human consumption or those sold to multiple companies. Thus, recorded byproduct levels may not be consistent with potential quantities because the data do not cover byproducts sold exclusively to one company for the production of a specialty product.

The data gathered from AMS can be used to estimate the total volume of beef and pork variety meats produced in the United States (see fig. 1). AMS data are the only production numbers that reflect actual counts and not estimates. This information, however, is provided voluntarily and may therefore represent only a portion of the byproducts harvested from hogs and cattle.

Another way to determine the volume of animal byproduct production is to create an "upper-bound" estimate using available data sources. A comparison of byproduct quantities reported voluntarily and byproduct quantities estimated by ERS provides evidence of the discrepancy between potential and reported amounts in the manufacturing sector. Ockerman and Hansen (2000) suggest that animal byproduct volumes can be estimated from slaughter numbers and dressing percentages.

In addition to shrinkage losses, more than 2 percent of the carcass weight is often unaccounted for (Ockerman and Hansen, 2000). Confidential sources in academia and the packing and processing sectors reveal that byproducts collected from livestock can vary from animal to animal, packer to packer, and over time, making it difficult to determine the exact byproduct amount produced at any time.

ERS estimates are based on data on the number of head slaughtered for each respective species from USDA, National Agricultural Statistics Service's *Livestock Slaughter* and assessments gathered from scholarly literature on animal byproducts.

By subtracting both the weight of muscle cuts and trimmings sold at the retail level and the estimated shrink and other waste loss (stomach contents, water, etc.) from the live weight of the animal, it is possible to estimate an upper bound for potential byproduct production. It is also worth noting that some fat and bone that typically remain on the carcass at slaughter and are later removed also contribute to byproduct production.

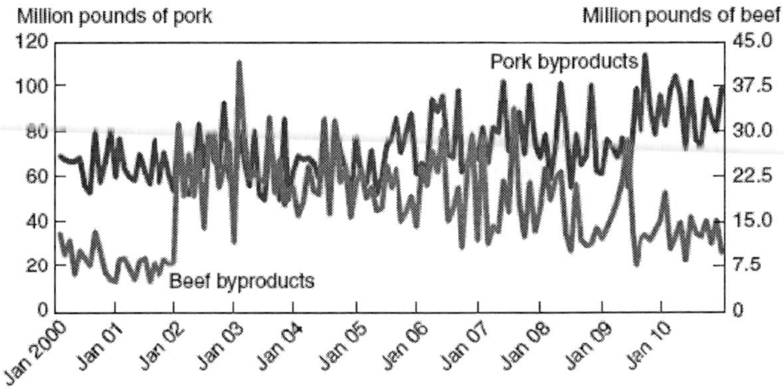

Note: Data are calculated by multiplying the total number of loads of delivered offal and byproducts of beef or pork recorded by AMS by 40,000 pounds, the weight of a standard load.Source: USDA, Economic Research Service using data from USDA, Agricultural Marketing Service (AMS).

Figure 1. Reported U.S. monthly beef and pork byproduct production from AMS data.

These estimates may be conservative because the numbers do not include condemnations that are rendered and added to the inedible byproducts market. Data from the literature reveal that about 11 percent of a typical slaughter hog's liveweight is lost through shrinkage or waste or is somehow unaccounted for (Aberle et al., 2001). After fat and bone have been trimmed off, about 77-79 percent of a dressed hog carcass weight produces muscle meat cuts and trimmings sold at the retail level (Aberle et al., 2001; USDA, ERS, 1992). The remaining share serves as an estimate of the total byproducts (volume) produced per hog. For cattle, about 14.1 percent of a typical steer's liveweight is lost through shrinkage or waste or is otherwise unaccounted for (Aberle et al., 2001; Duewer, 1998). After fat and bone have been trimmed off, about 69-70 percent of a dressed steer carcass weight produces muscle meat cuts and trimmings sold at the retail level. The remaining share serves as an estimate of the total volume of byproducts produced per steer.

Time series data reveal that total annual production of beef byproducts has remained fairly constant over the last decade or so, starting at about 18.5 billion pounds in 1997 and growing to almost 19 billion pounds in 2010 (fig. 2). Although estimates in other studies may include veal byproducts with beef byproducts, this analysis does not because, at most, veal byproducts add only one-half of 1 percent to total volume of beef offal production. The same time series data show that U.S. production of pork byproducts has grown over the last decade, much like the pork industry itself. Annual production of pork byproducts rose from about 7.1 billion pounds in 1997 to about 8.9 billion pounds in 2010.

Beef byproduct production is highly correlated with beef production (see figure 3). In 1921, only 38 percent of the liveweight of federally inspected cattle was sold at retail as muscle meat cuts and ground meat, and 48 percent of the animals' weight was available for production as beef byproducts. By 2010, these percentages were nearly equal: 42 percent of the animals' live-weight was used to produce retail muscle and ground meats, and 44 percent was available for byproduct production. This change is attributed to a shift in cattle genetics motivated by consumer preferences for more meat and less fat.

Pork meat production and pork byproduct production are also highly correlated. In 1921, 44 percent of the liveweight of federally inspected hogs was used as muscle cuts and trimmings meat and about 45 percent was available for production as pork byproducts.

Historical data on dressed weights from USDA's National Agricultural Statistics Service included back fat in the carcass weight until the late 1950s.

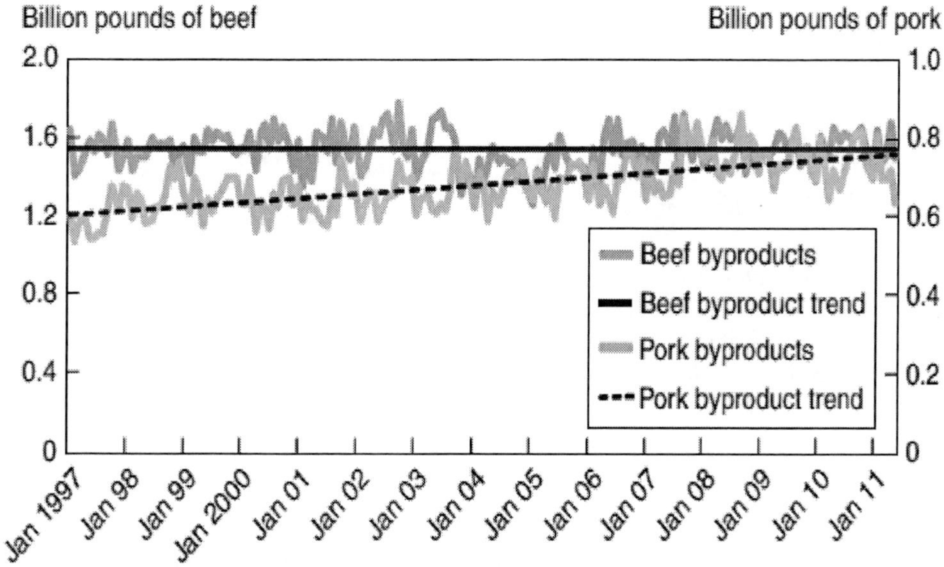

Source: USDA, Economic Research Service compiled from USDA, National Agricultural Statistics Service; Ockerman and Hansen, 2000; and Aberle et al., 2001.

Figure 2. Estimated monthly red meat byproduct production.

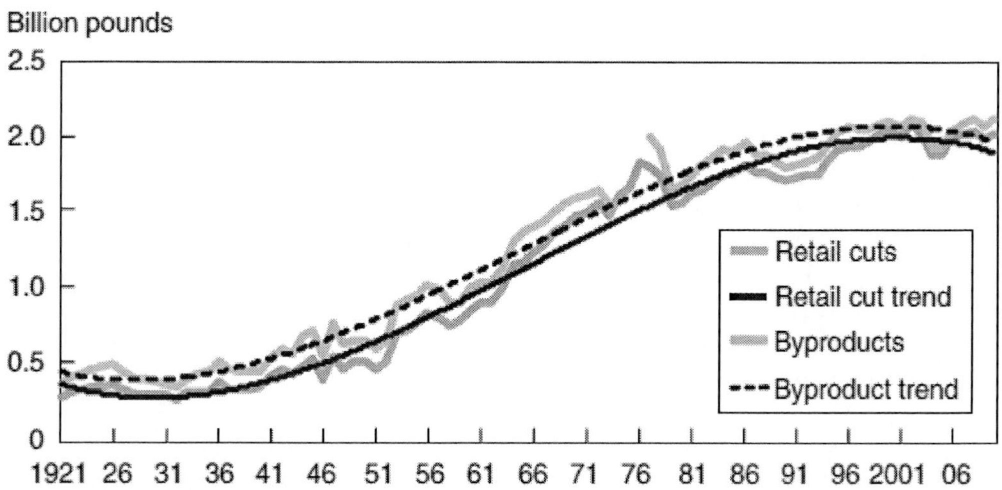

Source: USDA, Economic Research Service using data from USDA, National Agricultural Statistics Service; Ockerman and Hansen, 2000; and Aberle et al., 2001.

Note: Due to gaps in live weight data from 1974-76, we were unable to provide a continuous set of byproduct production data.

Figure 3. Estimated beef byproduct production.

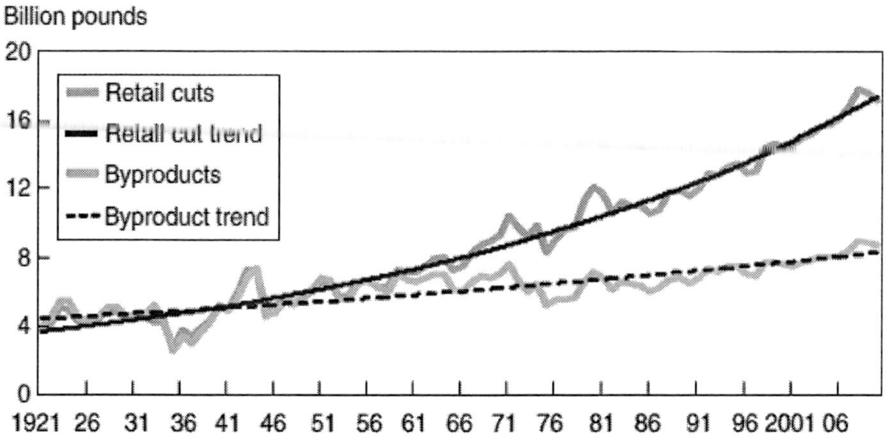

Source: USDA, Economic Research Service using data from USDA, National Agricultural Statistics Service; Ockerman and Hansen, 2000; and Aberle et al., 2001; and Ron Plain, University of Missouri.

Figure 4. Estimated pork byproduct production.

Therefore, this analysis adjusts the data to extract fat from dressed weights prior to their separation in the USDA data to reflect a more homogenous dataset (Plain, 2010) (see fig. 4). As hog fat and lard became less valuable beginning in the 1960s, the hog industry began breeding animals to meet the evolving demand for leaner animals with more muscle content. By 2010, 59 percent of the hog's weight produced retail muscle and trimming meat and only 30 percent was available for byproduct production.

Byproduct Values

Converting animal byproducts to useful goods with value has two effects. First, the meat industry captures additional revenue that otherwise would have been unrealized. Second, the costs of disposing of these secondary items are avoided.

Intrinsically, animal byproducts serve as an extra means for packers to earn revenue or as a cushion to cover losses should the cost of purchasing the live animal exceed the selling price of the carcass.

In the last decade, live-equivalent hog prices have risen faster than carcass cutout values, which are the prices received by the packer.

The hog carcass cutout value is an estimate of the value of a 51-52-percent lean, 185-pound hog carcass, based on wholesale prices being paid for subprimal pork cuts obtained from primal cuts, the basic major cuts into which carcasses and sides are separated. Pork primals include the loin, butt, picnic, sparerib, ham, belly, and jowl.

From 2000 to 2010, annual hog carcass cutout values rose by 27.7 percent (fig. 5). During the same period, however, the annual hog byproduct drop value increased by 80.3 percent.

Currently, 2011 has had record-high year-to-date hog byproduct drop values. This allows packers to bid higher prices for live hogs than they might have in the absence of the value added from byproducts.

The importance of byproducts to livestock production can also be assessed by examining the proportion of value derived from wholesale byproduct sales in relation to the value generated from sales of total wholesale items produced from the animal.

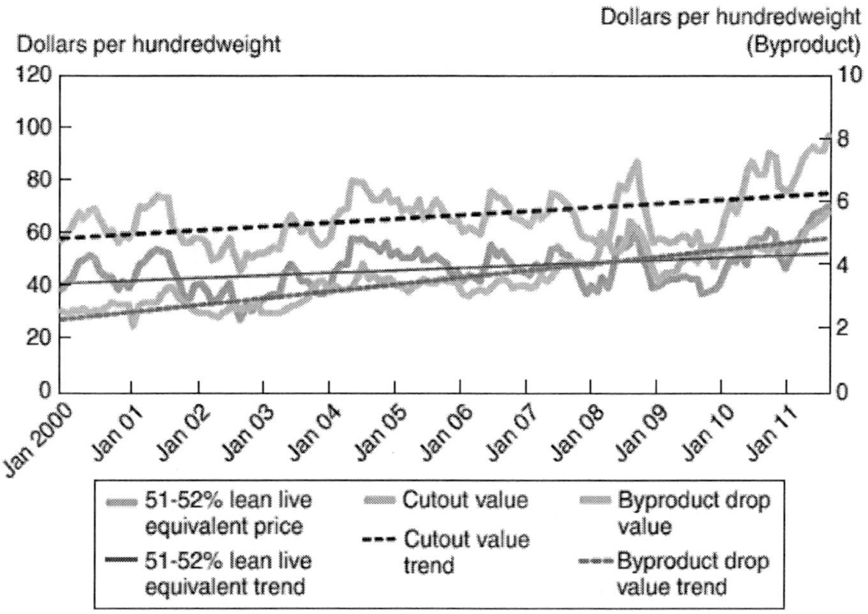

Source: USDA, Economic Research Service using data from USDA, Agricultural Marketing Service.

Figure 5. Prices and trends of U.S. live hog, carcass cutout, and byproduct values.

On a dressed-value basis, the total wholesale value of an animal can be defined as:

$$(TWv) = (COv) \times (COdw) + (BPv) \times (BPlw),$$

where ***TWv*** is the total wholesale value of an animal, ***COv*** is the weighted average cutout value of an animal's dressed carcass, ***COdw*** is the cutout dressed weight of the animal's carcass, ***BPv*** is the byproduct drop value per live weight of an animal, and ***BPlw*** is the live weight of the animal. Therefore, the portion of the animal's value derived from the byproduct value is at least:

$$\frac{BPv * BPlw}{Twv}$$

This calculation may not reflect the full value of the animal derived from byproduct value because kidney fat (diaphragm fat) and certain bones trimmed off of the carcass and sold as byproducts are still attached when the dressed carcass is weighed. Therefore, one can say that *at least* this much value is added from byproducts. As a greater percentage of the total value gained from the animal is derived from the byproduct value, byproducts are becoming increasingly more valuable to livestock feeders, as packers are willing to bid higher prices for these animals. For 2000, an estimated 5.1 percent of annual packer earnings per hog came from byproduct sales.

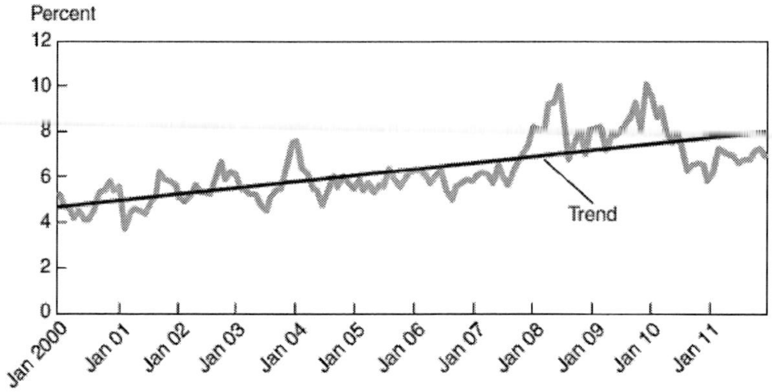

Source: USDA, Economic Research Service using data from USDA, Agricultural Marketing Service.

Figure 6. Percent of value added to hogs from byproducts.

Although down from its high of 8.7 percent in 2009, the annual proportion of packer revenue earned per animal from byproducts was 7.0 percent in 2010, 36.7 percent above what was earned in 2000, which indicates that hog byproducts have grown in importance in packers' margins over time (fig. 6). This growth has important implications in the context of marginal profitability. Likewise, as the value of byproducts increases, all else remaining equal, the total value of live animals also increases. Numerous factors help account for the increase in byproduct values in recent years. Historically, byproducts were once more commonly consumed during meals, in addition to being used for a number of household products. Over time, cheaper synthetic materials were substituted in household products and labor and other costs associated with production and harvesting of byproducts rose, resulting in a decline in byproduct values.

Further, U.S. immigrants and their descendants increasingly developed preferences for muscle cuts and other meat products, which also reduced demand for byproducts. More recently, however, a wider variety of uses and outlets for offal products has raised the value of these products relative to that of muscle cuts. The pet food industry, for example, has significantly boosted demand and, hence, value for animal byproducts. Common items found in retail store pet stores now include rawhide "bones," pizzles, joints and tendons, dry pet food made from processed offal, and pig ears, which are one of the most expensive parts of a hog carcass on a per weight basis.

Steer byproduct values, like those for pork, have increased since 2000, although not as significantly as pork byproducts. A sharp rise in foreign demand for swine offal in recent years, countered by a decline in demand for beef/veal offal, may account for the difference in growth rates. This may be because foreign demand for swine offal has risen quickly while bovine offal exports have fallen over the last few years, as discussed later in the trade section of this paper. The annual drop value of steer byproducts rose by 34.8 percent from 2000 to 2010, and similar to that for pork, beef carcass cutout prices also trended upward over the period. This study uses a weighted-average cutout value of Choice and Select cattle prices as an indicator for the prices received by the packer. This value simply accounts for the proportion of cattle slaughtered in each grade (Choice or better and Select) by the respective cutout value for each: a price calculated from the value of beef primal cuts, including ribs, chuck, round, loin, brisket, short plank, and flank. The weighted-average cutout value increased 36.4

percent from 2000 to 2010, an indication that packers are receiving, on average, over 36 percent more per carcass than they received a decade ago (fig. 7).

However, the five-area live steer price[2] trended upward by 38.1 percent during the same time period.

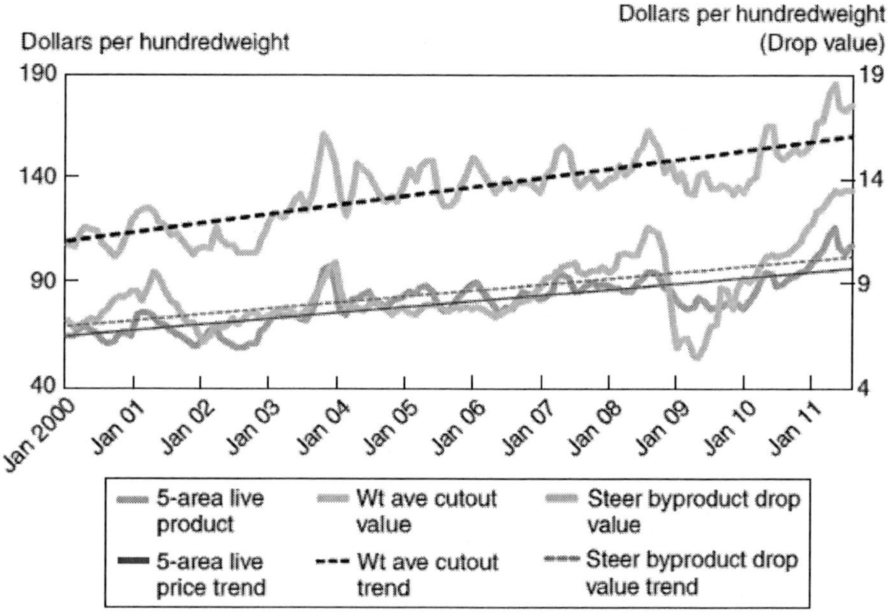

Source: USDA, Economic Research Service using data from USDA, Agricultural Marketing Service.

Figure 7. Prices and trends of U.S. live steer, beef carcass cutout, and byproduct values.

Although it may seem counterintuitive that the live price rose faster than the cutout price given the drop value's slower growth, a 36.4-percent increase in the greater valued cutout price amounts to more on a per head basis than a 38.1-percent increase in the lesser valued live price. In this instance, a 41-cents-per-pound increase for weighted average cutout price is compared with a 26-cents-per-pound increase for the five-area live price. As with hog byproduct drop values, steer byproduct drop values for 2011 have also seen record high levels year-to-date.

As shown in figure 8, the annual average proportion of value added to a steer from byproducts remained virtually the same from 2000 to 2010, at 10.3 percent. However, from January to September 2011, the proportion averaged over 11 percent.

Since a large share of byproduct value for cattle comes from hides, these values tend to follow hide values closely. For example, at the end of 2008, hide values plunged as effects of the worldwide recession dampened auto industry demand for leather and leather products.

Several factors help account for the sluggish growth in the value of beef byproducts relative to the value of the whole animal. Efficiency gains from technological advances have lowered the costs of recovering byproducts, enabling packers and renderers to sell more byproducts at a lower price and still maintain profitability.

Also, technologies now exist for producing synthetic materials that replace byproducts, including synthetic materials to make items that were once made of leather, vegetable fats and oils to replace animal fats in cooking, synthetic detergents to replace soaps formerly made from animal fat, and cellulose sausage casings to replace casings made from intestines (Aberle et al., 2001).

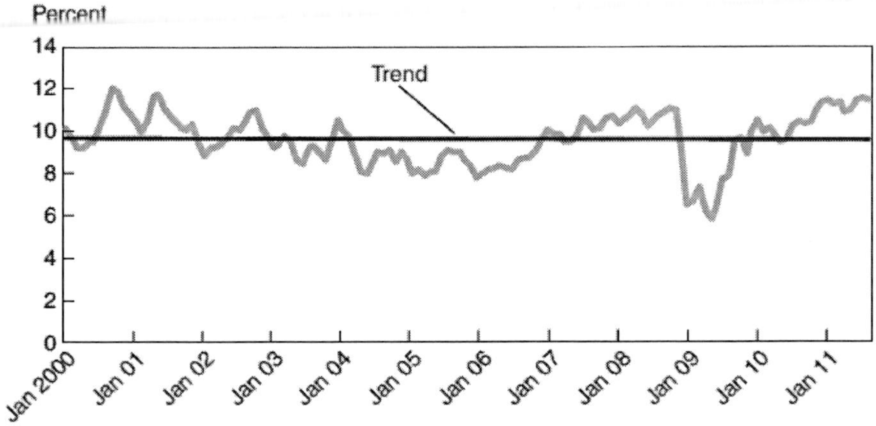

Source: USDA, Economic Research Service using data from USDA, Agricultural Marketing Service.

Figure 8. Percent of value added to the steer from byproducts.

In addition, BSE affected byproduct markets in a number of ways and continues to restrict beef and beef/veal byproduct exports and export growth. For example, many countries imposed restrictions on imports of U.S. beef and variety meats: Mexico and South Korea banned imports of small intestines, China banned imports of beef and variety meats, Hong Kong banned imports of beef from cattle over 30 months of age, and Japan banned imports of beef and products from cattle over 21 months of age. Also, some products historically categorized as "edible" in certain countries are now classified as "inedible" (brains, spinal tissue, spinal cords, etc.) and, in some cases, are considered a biohazard.

Empirical Example of Determining the Contribution of Byproducts to Steer Prices

To gain a greater understanding of how byproducts add to the value of a steer, ERS developed ad hoc regression model estimates of the monthly weighted average, five-area steer prices for all grades of steers for the period January 2000 through July 2010. Model variables are outlined in appendix B. The estimate for the cattle byproduct drop value is positive and statistically significant, indicating that byproducts contribute positively to the value of live cattle. These findings suggest that a 10-percent increase in the byproduct drop value will result in a 1.1-percent increase in the five-area weighted average price for all grades of steers. For a full description of the methodology and findings of this model, see appendix B.

The carcass cutout value is statistically significant and positively contributes to live cattle value, whereas the number of head slaughtered is also statistically significant but negatively correlates with cattle prices. This indicates that shortages support cattle prices while surpluses depress them. The model shows that the effect of the BSE outbreak in Japan on cattle prices is negative but not significant. In fact, when BSE was discovered in Japan in September 2001, Japanese demand for beef fell, regardless of beef origin, which reduced foreign demand for

U.S. beef and depressed steer prices. The effect on steer prices of the discovery of BSE in Canada is also significant and positive. This implies that when a case of BSE was found in Canada in May 2003, bans against imports of Canadian cattle affected U.S. domestic beef supplies, putting upward pressure on U.S. beef prices. Conversely, the effect of the U.S. BSE outbreak is negative, reflecting the decline in demand for U.S. beef when a case of BSE was found in a U.S. cow previously imported from Canada in December 2003.

FOREIGN BYPRODUCT PRICE COMPARISON

To gain insight into the value of byproducts for U.S. packers and renderers, it is also useful to observe prices for the separate items that comprise the edible byproduct in both domestic and foreign markets. Tables 1 and 2 illustrate the difference in value for select products available in U.S., Chinese, and Mexican markets. All U.S. prices are weighted averages taken from the USDA, AMS *Pork and Beef Variety Meats Report*, and all foreign prices are converted to U.S. dollars.

Table 1. U.S.-Chinese-Mexican pork byproduct price comparison

Product	U.S. average price,2000-2010	U.S. average price, 2010	U.S. price, week ending May 22, 2010	China price, May 20, 2010[1]	Central Mexico price, May 21, 2010[2]
			Dollars per hundredweight		
Kidneys	19.44	21.71	NA	146.11	NA
Livers	17.23	16.73	NA	51.80	NA
Ears	97.09	134.57	139.00	162.71	NA
Stomachs	65.23	106.85	112.00	112.90	NA
Chitterlings	42.41	NA	NA	83.01	NA
Hocks	29.06	42.73	37.00	69.73	NA
Hearts	33.84	32.47	33.00	89.66	NA
Tails	43.43	66.69	63.80	212.52	NA
Feet	34.97	46.85	51.00	83.01	57.69
Skins	25.00	40.00	NA	53.13	69.92
Tongues	75.97	135.30	122.00	136.14	NA
Head	NA	NA	NA	68.00	54.19
Cheek meat	66.17	93.48	94.70	NA	97.89
Visceras[3]	NA	NA	NA	NA	20.98

Source: USDA, Economic Research Service using data from USDA, Agricultural Marketing Service, LGMN Portal Database; Beijing Xinfadi Wholesale Market "price quotations" (translated by Fred Gale, ERS); and Sistema Nacional de Informaciÿn e Intergraciÿn de Mercados.

NA: Not applicable because these markets have not reported prices for these particular items during this time period.

[1] Chinese prices in terms of U.S. dollars use the exchange rate of 6.83 yuan to 1 dollar, as recorded by the Board of Governors of the Federal Reserve.

[2] Mexican prices in terms of U.S. dollars use the exchange rate of 12.974 pesos to 1 dollar, as recorded by the University of British Columbia Pacific Exchange Rate Service.

[3] An aggregated price for liver, heart, stomach, intestines, lungs, kidneys, and spleen sold in Aguascalientes, Mexico.

Table 2. U.S.-Chinese-Mexican beef byproduct price comparison

Product	U.S. average price, 2000-2010	U.S. average price, 2010	U.S. price, week ending May 22, 2010	China price, May 20, 2010[1]	Central Mexico price, May 21, 2010[2]
			Dollars per hundredweight		
Kidneys	19.30	30.81	NA	53.24	NA
Tripe	43.59	49.06	NA	99.83	NA
Liver	25.89	54.16	NA	53.24	NA
Heart	28.91	47.08	NA	66.55	NA
Tongue	154.45	203.40	NA	133.10	NA
Visceras[3]	NA	NA	NA	NA	43.72

Source: USDA, Economic Research Service (ERS) using data from USDA, Agricultural Marketing Service, LGMN Portal Database; Beijing Xinfadi Wholesale Market "price quotations" (translated by Fred Gale, ERS); and Sistema Nacional de Información e Intergración de Mercados.

NA: Not applicable because these markets have not reported prices for these particular items during this time period.

[1] Chinese prices in terms of U.S. dollars use the exchange rate of 6.83 yuan to 1 dollar, as recorded by the Board of Governors of the Federal Reserve.

[2] Mexican prices in terms of U.S. dollars use the exchange rate of 12.974 pesos to 1 dollar, as recorded by the University of British Columbia Pacific Exchange Rate Service.

[3] An aggregated price for liver, heart, stomach, intestines, lungs, kidneys, and spleen sold in Aguascalientes, Mexico.

Recent annual average prices for most of the variety meats are much higher than their 10-year averages, indicating that prices for these products are on the rise overall. Prices in the United States for some items, like stomachs, have trended upward in value closer to values in foreign markets. Greater international demand and higher prices for byproduct items add value to the live U.S. animal. As shown in the tables, certain U.S. items are particularly competitive in the Chinese and Mexican markets. While transportation costs, exchange rates, and tariffs all affect the costs of byproduct items imported from the United States, certain goods, such as kidneys, livers, hearts, tails, and tripe, cost more than twice as much in China than in the United States. Prices for these goods in Mexico, the closer of the two foreign markets, fall in between those in the United States and China.

U.S. Edible Offal and the Global Marketplace

The supply of offal produced in the United States is much larger than its domestic demand. U.S. demand for edible offal stems mostly from use as ingredients in products such as sausages, hot dogs, and pet foods. Variety meats are more highly valued in foreign markets than in the United States. Exporting U.S. edible offal to markets with greater preferences for the products increases the overall value and profitability of the carcasses (Reed and Saghaian, 2004). Although carcasses and high-value cuts account for the majority of both volume and value of total U.S. red meat exports, edible offal exports constitute about 26 percent of the volume and about 15 percent of the value of total beef/veal- and pork-product exports over the last 5 years. The United States was historically the world's largest exporter of edible offal up until 2003. Although the United States is still the single largest exporter of edible offal, in 2004, the EU-27 as a region overtook the United States as the world's largest export trader of these products. Data from 2010 suggest that the United States supplies about 28 percent of the world's exports of edible offal for all species; the EU supplies about 36 percent (fig. 9). Still, the U.S. share is much less than what it was in the 1950s, when the United States accounted for 32 percent of world exports of edible offal (USDA-FAS, 1958).

Pork Edible Offal Exports and Markets

Exports of U.S. pork edible offal have steadily increased over the last two decades (fig. 10). The percentage of pork products exported from edible offal has averaged 22.5 percent annually over the last 5 years, as exports of U.S. pork edible offal have more than tripled from levels a decade earlier to almost 481,000 metric tons (MT)[3] in 2010. The 2010 exports were the largestquantity of pork offal exports to date, accounting for 25 percent of the total volume of U.S. pork exports, the largest percentage in 6 years. In recent years, the value of U.S. exports of pork edible offal has also risen to historical highs. Prior to 2004, this amount was less than $200 million annually; it grew to $349 million in 2004 and has since more than doubled, reaching $708 million in 2008. In 2010, U.S. exports of pork edible offal were valued at $700 million, the third highest year on record.

Mexico is the largest importer of U.S. pork edible offal, accounting for over 42 percent of the U.S. exports over the last decade (fig. 11). Other major importers of U.S. pork edible offal include Hong Kong/China, Russia, Japan, and South Korea; however, many of these markets have developed only in the last 2-3 years. In this study, Hong Kong and China are grouped as one export destination.

Hong Kong is more of an open market with fewer trade barriers, and as suggested by findings in another study, Hong Kong re-exports most of the pork edible offal imports to China (see Bean, 1996). In 2008, Hong Kong/China began to rival Mexico as the leading export market for U.S. pork edible offal in terms of volume. Until 2004, exports of all U.S. pork edible offal to Hong Kong/China were marginal; in 2008, they jumped to over 130,000 MT, close to a third of the U.S. total shipped worldwide. In 2010, Hong Kong/China became the number one export destination for U.S. exports of pork edible offal. With imports of 182,000 MT, or nearly 38 percent of total U.S. shipments, Hong Kong/China set a record among all countries.

Prior to 2008, U.S. exports of pork edible offal to Russia were also minimal; however, in 2008, exports to Russia spiked at almost 61,000 MT, contributing to the near doubling of total U.S. exports of pork edible offal over the last few years of the decade. Offal can more easily be exported to Russia than muscle meat cuts because it does not fall under the tariff rate quotas that Russia imposes on other U.S. meats.

Major U.S. exports of pork edible offal in 2010 included fresh or chilled offal (13 percent of U.S. pork offal exports), hog feet (11 percent), rinds (8 percent), guts, bladders, and stomachs (8 percent), frozen intestines (7 percent) and all other frozen pork offal (45 percent) (fig. 12). Mexico imported 93 percent of all U.S.-exported pork rinds, 93 percent of the prepared or preserved offal exports, and 65 percent of the fresh or chilled pork offal exports in 2010. Russia was the leading importer of U.S. swine head meat, and Hong Kong/ China was the leading importer of U.S. hog feet, guts, stomachs and bladders, intestines, tongues, and pig hearts.

U.S. exports of beef and veal edible offal have been steadily increasing since being disrupted by the BSE outbreak in 2003. In 2010, bovine meat exports from the United States were 89 percent of their 2003 levels, and bovine edible offal exports were 74 percent of their 2003 levels.

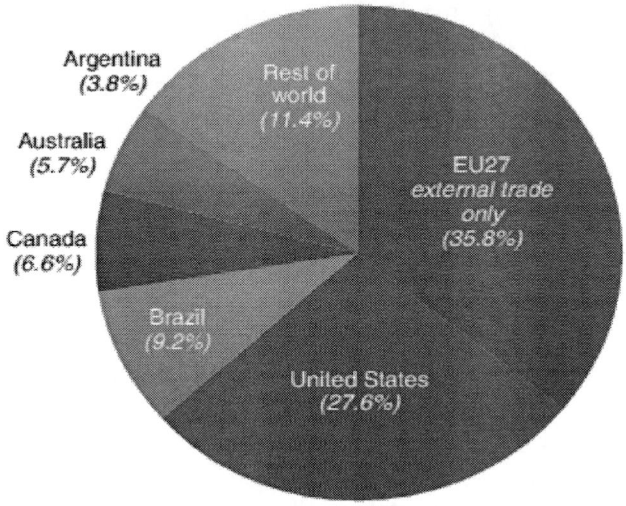

Figure 9. Major world edible offal exporting[1] countries by volume, 2010.

[1]. Data include offal from bovine, swine, goats, and horses (HTS code sets 0206) and guts, bladders, and stomachs of animals other than fish (HTS code set 0504). At least 90 percent of these offal are of bovine and swine origin.

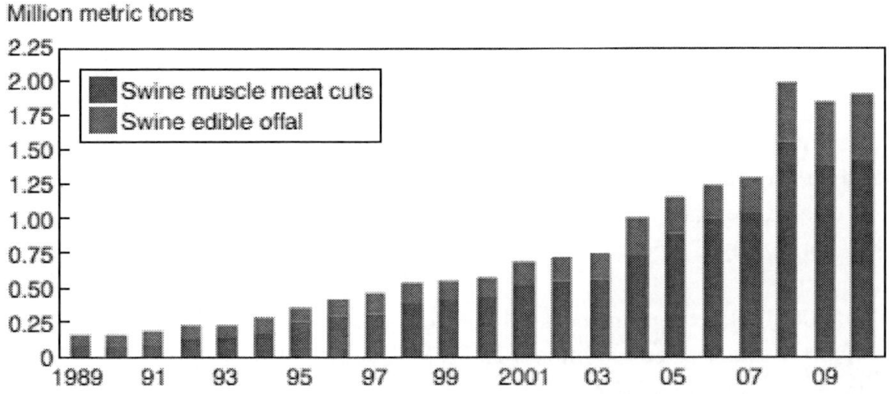

Source: USDA, Economic Research Service using data from USDA, Foreign Agricultural Service, Global Agricultural Trade System.

Figure 10. U.S. exports of swine meat and edible offal.[1]

[1] Data include swine edible offal (HTS code sets 020630, 02064, 020649, 0504000020, 0504000080, and 16024910) and fresh, chilled, frozen, or prepared pork (HTS code sets 0203, 0210, 160241, 160242, 16024920, 16024940, and 16024970).

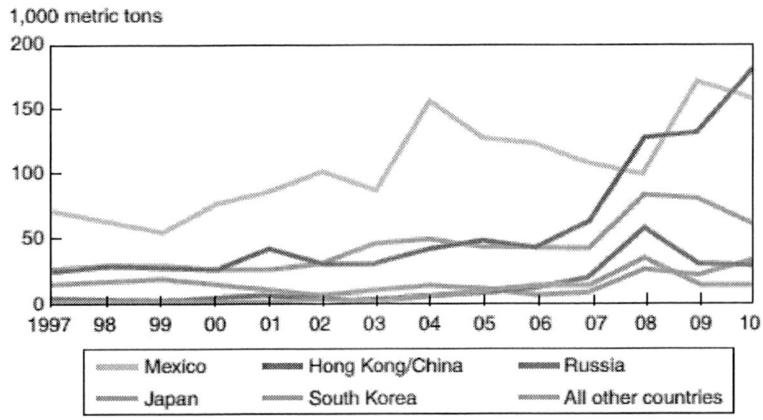

Source: USDA, Economic Research Service using data from USDA, Foreign Agricultural Service, Global Agricultural Trade Systems.

Figure 11. Top five export markets[1] for U.S. pork edible offal.

[1] These data include swine edible offal (HTS code sets 020630, 020641, 0504000020, 0504000080, and 16024910).

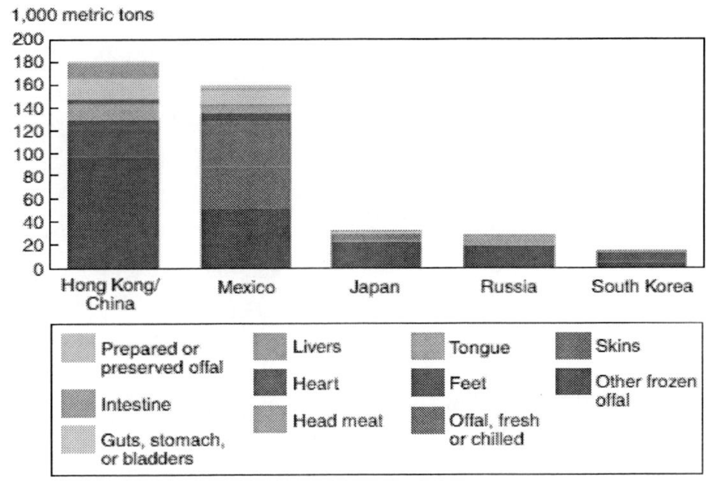

Source: USDA, Economic Research Service using data from USDA, Foreign Agricultural Service, Global Agricultural Trade System.

Figure 12. Top five export markets for U.S. pork edible offal by meat types, [1] 2010.

[1] These data include swine edible offal (HTS code sets 020630, 020649, 0504000020, 0504000080, and 16024910).

Beef Edible Offal Exports and Markets

Although edible offal's share of bovine product exports in the last 3 years has been less than the historical average, it reached almost 29 percent in 2009 (fig. 13). From 1990 to 2003, the average annual growth in the value of bovine edible offal exports was 10.7 percent, meaning that the value of bovine offal exported from the United States rose by over $413 million to $711 million during that period. After the initial decline of exports in 2004, the annual value of exports of U.S. bovine edible offal again climbed, reaching $541 million in 2010.

Historically, Japan was the largest importer of U.S. bovine edible offal, buying 32 percent of U.S. exports from 1997 to 2003. In Japan, items such as tongue, liver, stomach, and intestines are commonly used in barbecue, hot pot dishes, stews, and soups for both at-home and away-from-home consumption (Obara et al., 2010). Other traditionally large importers of U.S. bovine edible offal include Mexico, Egypt, Russia, and Canada (fig. 14). In these markets, "end cuts" and variety meats—the traditionally lower priced beef products in the United States—are bid away from competing uses in the United States. This contributes to an increase in the U.S. cutout value, and, ultimately, value is added to U.S. cattle at slaughter.

U.S. exports of bovine edible offal to major markets have increased gradually since 2003. During the same period and prior to the recent political turmoil in Egypt, U.S. exports of the product to Egypt also rose dramatically. In 2008, about 33 percent of U.S. exports of bovine edible offal went to Egypt, making it the largest destination for U.S. bovine variety meats that year (fig. 14). Having risen since 2006, U.S. exports to Egypt slowed in 2008-09 due in part to increased competition from Russia. Once Russia reopened its market to the United States, it started buying more U.S. beef livers, which drove up prices for Egyptian importers.

However, a larger factor in the 2008-09 slowdown may have been the global economic crisis, which resulted in tightened credit and a weakened Egyptian pound. In 2010, U.S. exports of bovine edible offal to Egypt almost recovered to their peak levels of 2007. The increase was driven by the efforts of Egyptian entrepreneurs who put beef liver and other price-competitive items in Egypt's retail outlets and supermarkets as well as on the menus of hotels and restaurants. Mexico currently remains the top export destination for U.S. bovine edible offal, if only by a small margin. In 2010, Mexico imported 87,700 MT of bovine edible offal from the United States, just 2,600 MT more than Egypt.

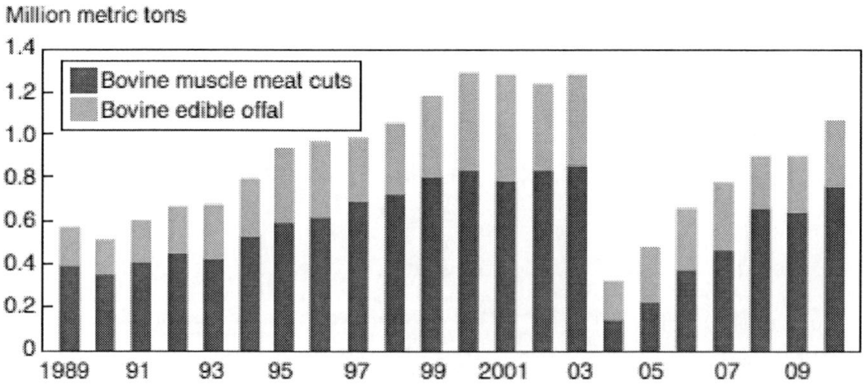

Source: USDA, Economic Research Service using data from USDA, Foreign Agricultural Service, Global Agricultural Trade System.

Figure 13. U.S. exports of bovine meat and edible offal. [1]

[1]. Data include bovine edible offal (HTS code sets 020610, 020622, 0504000050, 0504000070), and fresh, chilled, frozen, or prepared bovine meat (HTS code sets 0201, 0202, 16024920, 1602509020, and 1602509500).

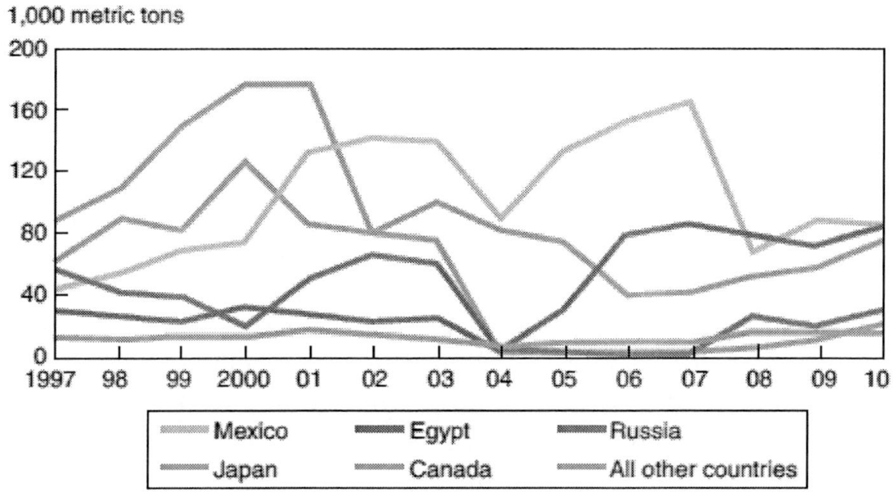

Source: USDA, Economic Research Service using data from USDA, Foreign Agricultural Service, Global Agricultural Trade System.

Figure 14. Top five export markets for U.S. bovine edible offal. [1]

[1] These data include bovine edible offal (HTS code sets 020610, 020622, 020629, 0504000050, 0504000070).

Traditionally, beef liver has been the top U.S. export of bovine edible offal. In 2010, bovine livers accounted for 31 percent of all U.S. exports of bovine edible offal (fig. 15). Other top U.S. exports of bovine edible offal include tripe (14 percent of U.S. edible offal exports) hearts (8 percent), tongue (5 percent), lips (4 percent), and all other frozen offal (29 percent). In 2010, 53 percent of all U.S. bovine liver exports went to Egypt, with another 21 percent going to Russia. Egypt also accounted for 35 percent of U.S. bovine kidney exports. In 2010, the majority of all U.S. exports of beef lips, sweetbreads, tripe, and tongue went to Mexico; more than half of U.S. exports of fresh or chilled bovine offal went to Japan; and about 28 percent of U.S. exports of all other frozen bovine offal went to Egypt, with Mexico taking the second highest volume.

Together, edible bovine and porcine byproducts accounted for more than 16 percent of the total value of U.S. bovine/porcine exports—which include edible offal and meat—over the last 10 years. In 2010, exports of pork and beef/veal edible offal accounted for only 14 percent of the total value of U.S. bovine/porcine exports, a lower-than-average share due to record-high export values of beef/veal and pork muscle meat. However, exports of beef/veal and pork edible offal in 2010 did reach a record $1.2 billion, $135 million more than the previous high set in 2009. The U.S. export value of edible beef/veal and pork offal has increased every year since 2004 (fig. 16).

Total world imports of edible offal have increased significantly in recent years, and emerging markets for U.S. byproducts are evident. Currently, Hong Kong/China, Mexico, and Russia import the greatest shares of total world exports of edible beef/veal and pork offal in terms of volume (fig. 17). Combined, these countries accounted for two-thirds of the world's edible offal imports over the last 10 years. Imports of beef/veal and pork edible offal by Hong Kong and China increased significantly during the period, mostly because of rising populations and swiftly growing incomes. Historically, the Philippines has not been a major importer of edible offal, but in 2010, the country almost doubled its annual total and overtook Japan as the world's fifth largest importer of edible offal.

Worldwide, the types of proteins consumed are often dependent upon peoples' income levels as well as cultural and religious practices. Increasing per capita incomes and rising GDP may have varied effects on consumption and trade of variety meats, depending on how the products are viewed in each country. In countries such as Egypt, where certain variety meats are popular, increasing wealth and GDP growth may result in increased U.S. variety meat exports. Egyptian demand for bovine products should remain strong in the coming years, especially since Egypt has a younger population, a relatively high rate of economic growth, and a limited capacity to expand domestic production (Kamenski, 2006). Variety meats are also highly valued in many Asian markets, where, in the long term, increases in U.S. meat and variety meat exports are anticipated. However, in portions of the Mexican and Russian markets, for example, where variety meats are less highly valued, variety meat consumption may give way to increasing consumption of muscle cuts as tastes and preferences change.

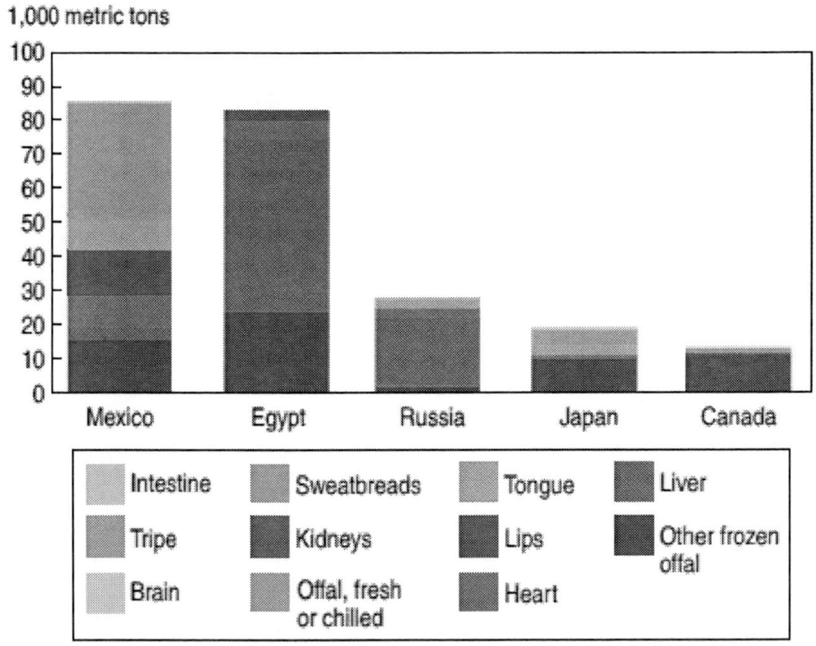

Source: USDA, Economic Research Service using data from USDA, Foreign Agricultural Service, Global Agricultural Trade System.

Figure 15. Top five export markets for U.S. bovine edible offal by meat types,[1] 2010.

[1] These data include bovine edible offal (HTS code sets 020610, 020621, 020629, 0504000050, 0504000070).

While U.S. exports of variety meats could decline as incomes rise in some countries, preferences in other countries, such as China, for certain culinary traditions that are strongly tied to variety meat use will continue to play an integral role in demand for U.S. variety meat exports.

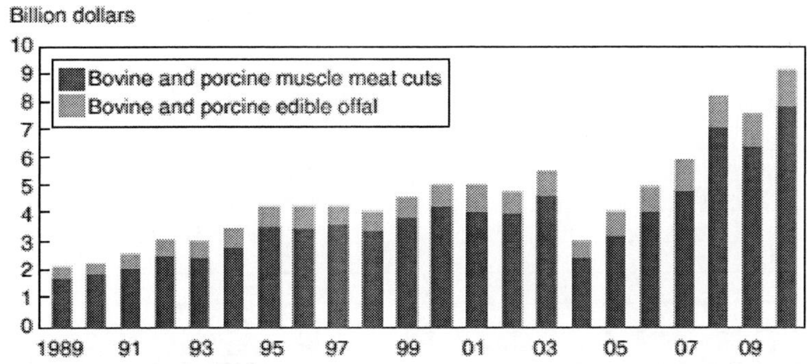

Source: USDA, Economic Research Service using data from USDA, Foreign Agricultural Service, Global Agricultural Trade System.

Figure 16. Export value of U.S. bovine and porcine muscle meat and edible offal. [1]

[1] Data include bovine and porcine edible offal (HTS code sets 020610, 020621, 020622, 020629, 020630, 020641, 020649, 0504000020, 0504000050, 0504000070, 0504000080, and 1602491000), and fresh, chilled, frozen, bovine and pork meat (HTS code sets 0201, 0202, 0203, 021011, 021012, 021019, 201020, 160241, 160242, 1602492000, 1602494000, 1602497000, and 160250).

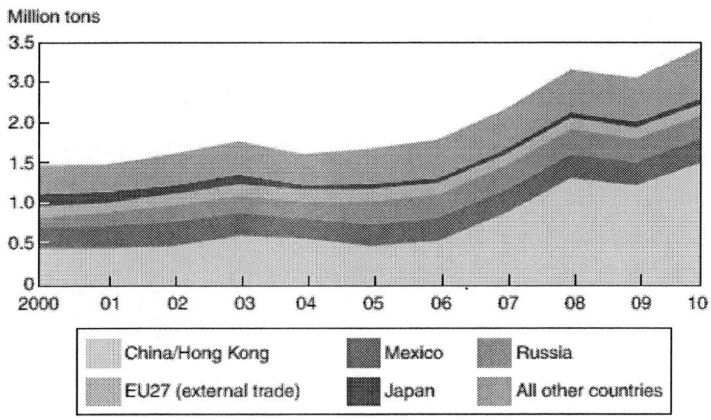

Source: USDA, Economic Research Service using data from Global Trade Information Services, Global Trade Atlas (complete data available only through 2009).

Figure 17. Historical top five world importers of edible offal by volume. [1]

[1] Data include edible offal from bovine, swine, sheep, goats, and horses (HTS code sets 0206) and guts, bladders, and stomachs or animals other than fish (HTS code set 0504). At least 80 percent of these offal are of bovine and swine origin.

CONCLUSION

The many uses of animal byproducts are often overlooked; however, these items contribute a significant value to the U.S. livestock and meat industries. Uses for animal byproducts include, but are not limited to, certain industrial, household, and cosmetic products; livestock feed additives; pet foods; pharmaceutical and medical supplies; human consumption; and foreign exportation.

APPENDIX A – TYPES OF OFFAL BY CATEGORY

Appendix Table A-1. Types of offal by category

Category	Raw byproduct	Principal use
Edible offal	Brain	Variety meats
	Head meat	Sausage ingredients, variety meats
	Heart	variety meats
	Kidneys	variety meats
	Liver	variety meats
	Spleen (melt)	variety meats
	Stomach (tripe)	Cheese making components, sausage components, variety meats
	Tail	Variety meats
	Testicles (fries)	Variety meats
	Thymus or pancreas (sweetbreads)	Variety meats
	Tongue	Variety meats
Inedible/edible offal	Blood	Adhesives, ceramics, cosmetics, feed use, fertilizer, foam in fire extinguishers, insecticides, laboratory use, medical use, plastics, sausage components
	Bones	Animal feed, buttons and handles, capsules for medications, cosmetics, emulsions, fertilizer, gelatins, glues, hardening steel, candies and dairy products, ointments, paper, photographic films, refining sugar, textiles
	Connective tissue	Gelatins
	Ears	Pet food, variety meats
	Fats	Candies, chewing gum, germicides, industrial oils, insecticides, lubricants, soap, glycerin, medicinal products, shortenings, tires
	Feet	Fine lubricants, leather preparations, variety meats
	Intestines (chitterlings or natural casings)	Sausage components, variety meats, medical use
	Skin	Candies, capsules for medications, cosmetics, dairy products, emulsions, gelatins, leather goods, ointments, paper, photographic films, rinds, textiles
Inedible offal	Glands	Industrial products, medicines
	Hair	Athletic equipment, brushes, felt, insulation, rugs, upholstery
	Hide	Boxes and plywood, gelatins, glues, leather goods
	Lungs	Pet foods

Source: USDA, Economic Research Service using data from Aberle et al., 2001; Ockerman and Hansen, 2000; Pearl, 2005; Goodlight, 2010; Corbin, 1992; and Murray et al., 1997.

Although the development of synthetic substitutes in the middle of the 20th century decreased the value of many animal byproducts, recent use in the pet food industry and the medical field and rising export demand are contributing to an increase in byproduct values in recent years. Evidence of byproduct value appreciation can be seen in both the beef and pork industries, but only for pork has the proportion of value added to the hog from the byproduct value increased significantly over the last decade. Empirical analysis reveals that wholesale byproduct values marginally influence the live animal's value. This finding is supported by the willingness of packers to pay marginally more (or less) for live animals based on current byproduct market values. Because the U.S. supply of edible offal typically exceeds domestic demand, the United States has historically been the world leader in exports of bovine and swine byproducts. Therefore, emerging markets that value offal products and are expanding their consumption will continue to play an important role in U.S. offal exports and thus, the U.S. livestock industry.

APPENDIX B — STEER BYPRODUCTS MODEL

To gain a greater understanding of how byproducts contribute to the value of a steer, we estimated an ad hoc regression model of the monthly weighted average, five-area steer prices for all grades of steers for January 2000 through December 2011.

Appendix Table B-1. Regression of weighted average five-area steer prices onto byproduct values and selected other variables, January 2000-July 2010

Variable	Description	Coefficient	Significance	t-Statistic
LOG(PR5AREASTEER)	Dependent variable: natural logarithm of 5-area weighted average price for all grades of steers sold on a negotiated basis, $/cwt			
C	Intercept	1.214785		1.763885
LOG(PRCHSELCUTOUT)	natural logarithm of weighted average Choice and Select cutout values, $/cwt	0.774930	***	12.32470
LOG(PRBYPROD)	natural logarithm of byproduct (drop) value, $/cwt	0.113227	***	5.094476
LOG(PR50TRIM)	natural logarithm of 50-percent lean trim, $/cwt	0.003789		0.297329
LOG(PRHOGS)	natural logarithm of prices for slaughter barrows and gilts, $/cwt	0.030727		1.044437
LOG(QFISLAU)	natural logarithm of Federally inspected cattle slaughter, million head	-0.096641	**	-2.365011
LOG(CATIMP)	natural logarithm of live cattle imports, million head	0.035176	***	4.859475
BSEJAPAN	dummy variable: 0 = before September 2001, 1= September 2001 and beyond			-1.417485
BSEUS	dummy variable: 0 = before January 2004, 1= January 2004 and beyond			-0.202876
BSECAN	dummy variable: 0 = before May 2003, 1=May 2003 and beyond		***	2.680094
r-square = 0.962392 F-statistic = 232.2765				

Source: USDA, Economic Research Service.
Significance levels are *** = 1% and ** = 5%

The log-log model specification facilitates the interpretation of the parameters as response parameters or a form of elasticity.

Variables used in the regression model are outlined in appendix table 1. A weighted average of Choice and Select cutout values is calculated to represent all grades. Cutout values, byproduct values, prices for 50-percent trim, and hog prices were originally from USDA, Agricultural Marketing Service reports. Data on total federally inspected cattle slaughter come from USDA, National Agricultural Statistics Service *Livestock Slaughter*. Data on live cattle imports are obtained from ERS web sources. Indicator variables for bovine spongiform encephalitis (BSE) confirmations in export markets important to the United States are constructed based on BSE confirmation dates. Trigonometric variables were added to account for seasonality and any effects from the cyclical nature of cattle inventories.

Any extra products from an animal would generally be expected to add value to the animal, meaning that they would have a positive effect on steer prices. This is the case in the model for cutout, byproduct, and trim values, although only the effects from cutout and byproduct values are statistically significant. The price of a substitute, hogs in this case, is also expected to be positive as it is in this model, but, again, not statistically significant.

Interpretation of the byproduct parameter—the main parameter of interest in this model—is that a 1-percent increase in byproduct values results in a 0.11- percent increase in steer prices. An increase in byproduct values from $13.20 to $13.33 (a 1-percent increase) implies an increase in the five-area steer price from $115.00 to $115.13 (a change in the steer price of 0.11 percent). Other parameters have similar interpretations but differ in magnitude and direction because of the signs attached to the parameter estimates.

Total federally inspected cattle slaughter, a quantity variable, was expected to have a negative effect on steer prices since, from a demand perspective, a larger quantity of the commodity would sell at a lower price. In this model, the sign for slaughter is negative and significant. The effect of a trade-related variable is not so easily explained. Imports of live cattle were expected to have a negative effect on U.S. steer prices because more cattle would sell at lower prices. However, the study is one in which cattle inventories were generally declining. Also, most of the imported cattle are feeder cattle, and a positive parameter may indicate demand for inputs into future production in the face of inadequate current supplies. The parameter for cattle imports is both positive and significant.

The indicator variable for BSE in Japan is negative but not significant, likely reflecting the adverse reaction of Japanese consumers to their own BSE events. The reluctance of Japanese consumers to consume any beef reduced exports of U.S. beef, and Japan was the top U.S. beef export market prior to their BSE confirmation. The U.S. BSE confirmation had an expected negative effect on steer prices but, interestingly, is not significant. U.S. prices for beef and cattle had been at record levels until October 2003 when they began declining. While initial responses were negative during the first couple weeks after the December 2003 U.S. BSE event, the effect was short lived, and U.S. consumers did not respond to the domestic BSE cases as adversely as consumers in other countries responded to their earlier BSE discoveries. The Canadian BSE variable was significant and had an expected positive effect on U.S. steer prices because reduced imports from Canada decreased domestic U.S. beef supplies, especially at a time when U.S. supplies were already low.

F-tests for seasonal and cyclical variables were not found to be significant. This was somewhat surprising because of the appearance of seasonality in the price series and results from other studies. However, seasonality exists in other variables, and it may be that the other variables captured the seasonal components, making seasonal variables redundant. The same is likely true for cyclical variables as well.

ACKNOWLEDGMENTS

The authors gratefully acknowledge the reviews of Erin Daley Borror, Thad Lively, Paul Clayton, Kevin Smith, and Courtney Heller of the U.S. Meat Export Federation; Claire Mezoughem and Lesley Ahmed of USDA's Foreign Agricultural Service; Sherry Wise, Warren Preston, Martin O'Connor, and Bucky Gwaltney of USDA's Agricultural Marketing Service; Herbert Ockerman of Ohio State University; Richard Stillman, William Hahn, Mildred Haley, and Molly Garber of USDA's Economic Research Service (ERS); and Shayle Shagam of USDA's World Agricultural Outlook Board. The authors would also like to acknowledge Fred Gale of ERS for gathering and translating Chinese pork prices and Ron Plain of the University of Missouri for providing historical carcass weight data. The authors appreciate and acknowledge the editorial and design assistance of John Weber and Victor B. Phillips, Jr. of ERS.

REFERENCES

Aberle, E.D., J.C. Forrest, D.E. Gerrard, and E.W. Mills. *Principles of Meat Science*, Fourth edition, Kendall/Hunt Publishing Co.: Dubuque, IA, 2001.

Bean, R. "Beef Tongues to Pork Maws, Foreign Markets Hunger for Variety Meats," *Ag Exporter*, June-July 1996, http://findarticles.com/p/articles/ mi_m3723/is_n5_v8/ai_ 1875 2623/?tag=content;col1

Beijing Xinfadi Wholesale Market. "Price Quotations," Translated by Fred Gale, U.S. Department of Agriculture, Economic Research Service, Accessed October 2010, www. xinfadi.com.cn/channel/13195963

Corbin, J.E. "Inedible Meat, Poultry and Fish By-Products in Pet Foods," in A.M. Pearson and T.R. Dutson (eds.), *Inedible Meat By-Products, Advances in Meat Research*, Vol. 8. pp, 329-347, Elsevier Applied Science: London, 1992.

Danilevici, C., M. Iordan, and M. Niă. "Valorisation of By-Products From Food Industry, Means of Effective Protecting the Environment," *Annals, Food Science and Technology*, 2009, www.afst.valahia.ro/anale2009/vol1/ sectiunea4/10_CONSTANTIN_ DANILE VI CI_2.pdf

Duewer, Larry. "Updated Beef Carcass-to-Retail Consumption Conversion Factor Increases to 0.7," *Livestock Dairy and Poultry Outlook*, Outlook Report No. LDP-51, U.S. Department of Agriculture, Economic

Research Service, May 1998, http://usda.mannlib.cornell.edu/usda/ers/ LDP-M/1990s /1998/ LDP-M-05-21-1998.txt

Dyck, J.H., and K.E. Nelson. *Structure of the Global Markets for Meat*, Agriculture Information Bulletin 785, U.S. Department of Agriculture, Economic Research Service, September 2003, www.ers.usda.gov/ publications/aib785/

Franco, D.A., and W. Swanson. *The Original Recyclers*, Animal Proteins Producers Industry, Fats and Proteins Research Foundation, and National Renderers Association, 1996.

Goodlight, A. "Medical Breakthroughs: Pig Parts for the Human Body," *Associated Content from Yahoo*, August 18, 2010, www.associated content.com/article/5684798/ medical_ breakthroughs _pig _parts_for. Html ?cat=58

Halstead, T. "U.S. Variety Meat Exports on the Upswing," *FAS online*, 1999, www.fas.usda. gov/dlp2/circular/1999/99-03LP/variety.htm

Hayes, Dermot. "Chinese Market for U.S. Pork Exports," *Trade Research Center*, August 1997, pg. 11, http://ageconsearch.umn.edu/ bitstream/ 29173/1/pip02.pdf

Kamenski, C. "Variety Meats Vital to Boosting U.S. Beef Exports," *Issues Update*, March-April 2006, pp. 29-30, www.beefusa.org/uDocs/varietymeatsvitaltoexports.pdf

Marti, Daniel L., and Rachel J. Johnson. "Special Article: U.S. Variety Meat Exports and the Global Marketplace," in *Livestock, Dairy, and Poultry Outlook*, Outlook Report No. LDP-M-195, U.S. Department of Agriculture, Economic Research Service, September 2010, www.ers.usda. gov/publications/ldp/2010/09sep/ldpm195.pdf

Mathews, K.H. Jr. "Economic Impacts of Feed-Related Regulatory

Responses to Bovine Spongiform Encephalopathy," in *Livestock, Dairy, and Poultry Outlook*, Outlook Report No. LDP-M-170-01, U.S. Department of Agriculture, Economic Research Service, September 2008, http://usda.mannlib.cornell.edu/usda/ers/LDP-M//2000s/2008/ LDP-M09-04-2008_Special_Report.pdf

Mathews, K.H. Jr., M. Vandeveer, and R.A. Gustafson. "An Economic Chronology of Bovine Spongiform Encephalopathy in North America," in *Livestock, Dairy, and Poultry Outlook*, Outlook Report No. LDP-M-143- 01, U.S. Department of Agriculture, Economic Research Service, June 2006. http://usda.mannlib.cornell.edu/usda/ers/LDP-M//2000s/ 2006/ LDP-M-06-09-2006_Special_Report.pdf

Mexico, Republic of, Secretaria de Economia. Sistema Nacional de Informacion e Integracion de Mercados website, Accessed October 2010, www.economia-sniim.gob.mx/Nuevo/

Murray, S.M., A.R. Patil, G.C. Fahey, Jr, N.R. Merchen, and D.M. Hughes. "Raw and Rendered Animal By-Products as Ingredients in Dog Diets," *Journal of Animal Science*, 1997, 75:2497-2505.

National Renderers Association. *Render*, Alexandria, VA, Various April issues.

Obara, K., M. McConnell, and J. Dyck. *Japan's Beef Market*, U.S.

Department of Agriculture, Economic Research Service, Outlook Report No. LDP-M-194-01, August 2010, www.ers.usda.gov/publications/ ldp/2010/08aug/ldpm19401/ldpm19401.pdf

Ockerman, H.W., and C.L. Hansen. *Animal Byproduct Processing and Utilization*, First edition, Lancaster, PA: Technomic, 2000.

Ockerman, H.W., and L. Basu. "Edible Rendering—Rendered Products for Human Use," April 29, 2010, https://dspace.lib.ohio-state.edu/dspace/ bitstream/1811/45284/1/ Rend ering_Chapter.pdf

Pearl, G.G. "Animal By-Products: Biological and Industrial Products," in W.G. Pond and A.W. Bell (eds.), *Encyclopedia of Animal Science* 19-21, Marcel Dekker: New York, 2005.

Plain, R.L. Professor of Agricultural Economics, University of Missouri, Personal communication, October 27, 2010.

Prokop, W. H. "The Rendering Industry—A Commitment to Public Service," in *The Original Recyclers*, D.A. Franco and W. Swanson (eds.), pp. 17-21, Animal Protein Producers Industry, Fats and Proteins Research Foundation, National Renderers Association, 1996.

Reed, M. R., and S.H. Saghaian. "Measuring the Intensity of Competition in the Japanese Beef Market," *Journal of Agricultural and Applied Economics* 36(1): 113-121, April 2004.

U.S. Department of Agriculture, Agricultural Marketing Service. "LGMN Portals Database," http://marketnews.usda.gov/portal/lg

U.S. Department of Agriculture, Agricultural Marketing Service. *Market News Reports*, http://www.ams.usda.gov/AMSv1.0/ams.fetchTemplateData.do?template=TemplateB&navID=MarketNewsAndTransportationData&leftNav=MarketNewsAndTransportationData&page=LSMarketNew sPage

U.S. Department of Agriculture, Economic Research Service. *Weights, Measures, and Conversion Factors for Agricultural Commodities and Their Products*, Agricultural Handbook No. 697, 1992.

U.S. Department of Agriculture, Foreign Agricultural Service (a). "Foreign Agricultural Trade of the United States," www.fas.usda.gov/ustrade/

U.S. Department of Agriculture, Foreign Agricultural Service (b). *United States Variety Meats in World Markets*, FAS M-43, October 1958.

U.S. Department of Commerce, U.S. Bureau of the Census. *Current Industrial Report, M311K—Fats and Oils, Production Consumption, and Stocks*, Various issues.

In: There's The Beef
Editors: Jonathan L. Ruiz and S. R. Fox

ISBN 978-1-61942-943-7

Chapter 4

JAPAN'S BEEF MARKET*

Kakuyu Obara, Michael McConnell and John Dyck

ABSTRACT

Japan imports large amounts of beef, primarily from Oceania and North America, and its consumers are willing to pay a premium for heavily marbled, grain-fed beef. As a result, Japan's import of certain beef cuts and offal raise the value of U.S. cattle. Trade bans resulting from the discovery of bovine spongiform encephalopathy (BSE) in North America dramatically shifted beef supplies to imported beef from Australia and New Zealand.

Beef consumption in Japan may increase from current levels in Japan's market, particularly if prices fall or income rises, despite a declining population. Economic factors, demographic factors, import and domestic policies and regulations, as well as consumer tastes and preferences, will determine the outlook for beef consumption in Japan and the ability of U.S. beef to compete in that market.

Japan's domestic beef production relies upon imported feed, primarily from the United States, to feed specific breeds energy-intensive rations. Government support plays an important role in the market structure of the domestic beef industry, which has an impact on imported beef from other countries, including the United States.

Keywords: Japan, beef, cattle, trade, beef markets, demand elasticity

ABOUT THE AUTHORS

Kakuyu Obara is an analyst with USDA's Foreign Agricultural Service. Michael McConnell and John Dyck are economists with USDA's Economic Research Service.

---

* This is an edited, reformatted and augmented version of the United States Department of Agriculture Publication, LDP-M-194-01, dated August 2010.

SUMMARY

Japan's beef consumption is important to U.S. agriculture. Beef eaten in Japan is either produced from Japanese cattle that are fed U.S. feedstuffs or imported. A large share of imported beef historically has come from U.S. cattle.

What Is the Issue?

Japanese beef consumption has been buffeted by severe shocks for over 15 years. As both a competitor of U.S. imported beef and a purchaser of U.S. feedstuffs, Japanese beef production has also suffered setbacks. As a result of the discovery of bovine spongiform encephalopathy (BSE), U.S. beef was completely banned from Japan's market in 2004 and has had only limited access since. In the aftermath of the BSE-related shocks that have affected consumption, production, and trade, a review of the entire Japanese beef market may provide insight into the potential for growth in Japanese beef consumption. It may also reveal the extent to which future Japanese beef production can satisfy a share of consumption and remain a market for U.S. feedstuffs and what the current and future role of U.S. beef might be in satisfying Japanese consumption.

What Did the Study Find?

Japanese beef consumption remains sensitive to income and price changes. Individual consumption of beef could rise significantly if income levels rise or beef prices fall. The price of U.S. beef, relative to substitutes from Australia or Japan, is an important determinant of U.S. trade share. If the price of U.S. beef drops, imports from the United States are likely to expand, perhaps even to the level reached in 2003 (when Japan was the largest foreign market for U.S. beef).

Domestic Japanese beef production has not filled the gap left by the restrictions on North American beef imports. High prices for feeder calves and high feed costs, together with a relatively small-scale feedlot industry, prevent Japanese production from increasing. Although strong Government subsidy support and a substantial import tariff continue to bolster the Japanese industry, beef production is unlikely to expand.

In recent years, Japan's beef cattle industry has intensified its feeding to increase certain beef attributes, such as marbling. This practice has helped support demand for U.S. corn and barley.

The increase in feed per head of cattle may not continue because Japanese consumers may be less enthusiastic, for health reasons, about heavily marbled beef. Beef production in Japan, however, is likely to remain one of the key foreign markets for U.S. feed grain.

Certain lower priced cuts of U.S. beef, primarily "end cuts" and offal, are considered a good value in Japan. The Japanese market bids these U.S. cuts away from competing uses, adding to the cut-out value of U.S. cattle at slaughter.

Current Japanese rules limit the supply of these cuts and raise their import prices. If these rules were changed, Japan's purchases of these cuts could increase, most likely at the expense of Australian beef, which currently has the largest market share of imported beef in Japan.

How Was the Study Conducted?

The report draws on several sources of information:

- Livestock and grain and feed reports for Japan from the Foreign Agricultural Service (USDA);
- Reports and data published by Japan's Ministry of Agriculture, Forestry and Fisheries (MAFF);
- Books and articles published about Japan's beef industry;
- Interviews with representatives of firms, Government officials, and academics involved in the beef industry in Japan (September 2008);
- Econometric analysis by the authors.

INTRODUCTION

Japan is the world's second leading beef importer by value, behind the United States, and once was the most important export market for U.S. beef. In 2009, Japan imported over $470 million in U.S. beef products (down from $1.6 billion in 2003 prior to trade restrictions). Japan's domestic production is a very important market for U.S. feedstuffs exports, with over $740 million imported to feed beef cattle in 2007 (the latest year with available data). Japan's beef consumption, still relatively low by developed-country standards, grew substantially in the first half of the 1990s after the liberalization of an import quota regime. Most of the new consumption was met by imports because domestic beef production in Japan is expensive.

A series of disease outbreaks and food safety incidents beginning in 1996 shook Japan's beef market. Demand was affected repeatedly and, sometimes, substantially. Discovery of bovine spongiform encephalopathy (BSE), the neurological disease also known as mad cow disease, in Japan's cattle herd in 2001 severely stressed domestic supply and depressed consumer confidence. Discovery of BSE in North America in 2003 initially halted beef imports into Japan from the United States and Canada and later was followed by restrictions on beef imports from North America.

These actions cut off a large share of Japan's beef supply, which has still not been replaced by other suppliers or restored from North American suppliers. Following 10 years of turbulence—1996 to the present—the future of Japan's beef market is still unclear.

Through most of the 1990s and into the new century, Japan's consumers and beef suppliers have operated in a stagnant economy. Income growth per person has been slight, although Japanese households remain among the wealthiest in the world. Population growth slowed over this period and became negative in 2005, a trend that is expected to continue. Given this environment, Japan may never consume as much beef as it did at its peak in 2000.

Japan's consumers have proven to be price sensitive. Imported beef prices fell as Japan reduced import tariff levels during 1991-2000 and, in some years, as Japan's currency strengthened. Before the removal of beef import quotas (1988-1991), experts speculated that Japan's wealthy consumers would seek out the most expensive imported cuts, but some inexpensive cuts from North American and Oceania beef production, in fact, have become best sellers. Japan's consumers are willing to pay more for these cuts (and some offal) than consumers in the exporting countries are, contributing to higher carcass prices for exporters. Despite the premium that they paid, Japanese customers regarded several popular imported cuts as good values in 2003 and before. Beef customers in Japan have a taste for specific cuts and qualities of meat that is different from that of consumers in the West. After trade bans were lifted in 2006, U.S. beef returned to Japan but at higher prices than in 2003, partly due to specific restrictions that limited supply. The higher prices indicate that demand for grain-fed beef products is still relatively high in Japan.

Japan sources beef both from both wagyu cattle—Japan's traditional draught animal—and Holstein cattle from the dairy herd. Japan's beef cattle farms, faced with a limited amount of land and feedstuffs, rely heavily on imported grains and roughage.

In recent years, more grain has been used per pound of beef produced to increase the marbling of the meat. As a result, Japan's beef production uses a significant share of U.S. corn exports. Japan also imports a significant amount of U.S. barley and other cereals, forages, and protein sources for feed rations. Thus, U.S. grain and forage supports both Japan's domestic beef and beef imported from the United States.

The Japanese Government, feed companies, and cooperative organizations play an important role in support and risk management for domestic cattle and beef producers. Several Government programs provide subsidies for producers, and an industry program helps mitigate fluctuating feed prices.

This report reviews the recent history and current situation of the Japanese beef market to help answer important questions about its future.

DEMAND FOR BEEF

Meat consumption in Japan is much lower than in many other countries (fig. 1). Large volumes of fish and seafood characterize the Japanese diet. Fish consumption, however, has stagnated in Japan over the past few decades, providing opportunities for growth in meat consumption. Meat consumption levels may never reach the levels in other societies, but Japan is a valuable market due to the demand for high-quality meat products with specific attributes. These trends are particularly true for the beef market.

Beef products in Japan are highly segmented. Beef is prepared differently depending not only on the type of cut, but also on whether the beef is wagyu or dairy, domestic or imported, grain-fed or grass-fed. In addition, income, price, consumers' ages, and preferences can change not only overall beef consumption, but also the type and way in which beef is consumed.

How Beef Is Eaten

Beef is eaten both at home and away from home. Japan's household consumption accounts for approximately 34 percent of beef consumption, while catering and food service accounts for 57 percent of consumption. Processing use—for use in cooked or otherwise preserved products—accounts for the final 9 percent (MAFF, 2007).[1] Cooking styles often influence what type of beef is preferred.

Traditional Japanese beef dishes are often prepared with short exposure to heat over an open flame or in boiling water. Because of this type of preparation, cuts are generally thinly cut and well marbled to attain the desired taste and texture. Examples of traditional beef dishes can be found at *yakiniku* and *shabu-shabu* restaurants. Yakiniku is a Korean-style grilled dish served with thin, raw strips of beef. The meat is cooked by the customer on an open-flame grill set up at the table. After cooking the meat for a few minutes, the meat is then eaten with other vegetables, rice, and soy items. Similarly, shabu-shabu restaurants also serve beef raw. The customer cooks the thinly sliced meat in boiling water or broth, and then eats it with different sauces. Other vegetables and noodles are also cooked in the boiling water. In both cases, thinly cut pieces of raw beef are used so that the meat is cooked evenly with a limited amount of heat. Heavily marbled beef is preferred as it provides a tender and flavorful taste and texture. In most cases, these dishes are prepared with domestic meat.

Gyudon, or beef bowl, is also a popular dish. Gyudon contains onions and beef simmered in sweet soy sauce and served over rice. It can also be served with pickled ginger and miso soup. Gyudon is relatively inexpensive and typically served as a fast-food item. A beef bowl restaurant is a popular destination for lunch among business people seeking a quick and inexpensive meal. Yoshinoya and Sukiya are the two largest gyudon chains in Japan. Prior to the discovery of BSE in North America, these chains used imported U.S. beef, primarily short plate, and were hurt by the trade bans initiated in 2003. While Sukiya switched to Australian beef, Yoshinoya stopped serving gyudon altogether. Instead, Yoshinoya introduced *buta don*, or pork bowl, which has grown in popularity. Yoshinoya resumed its gyudon in 2006, when U.S. beef was available in Japan again, but kept buta don on its menu. With limited supplies of U.S. short-plate due to age restrictions of cattle, brisket is now also used as a substitute. Even for this low-priced dish, the Japanese prefer grain-fed beef. In general, across the spectrum of the different beef dishes, an affinity for well-marbled beef means that grain-fed beef is used more often than grass-fed beef.

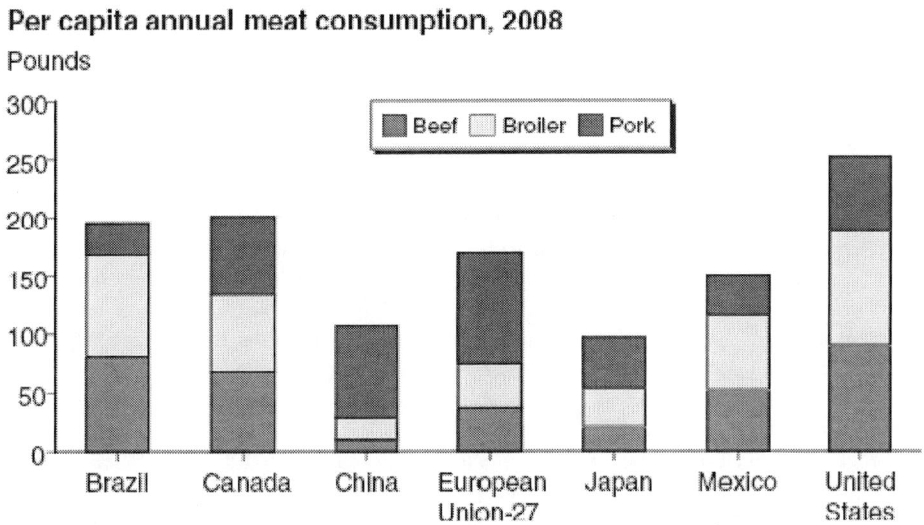

Source: USDA, Economic Research Service calculations based on USDA, Foreign Agricultural Service Production, Supply and Distribution Online database and population data.

Figure 1. Per capita annual meat consumption, 2008.

Hamburgers are another major preparation of beef in Japan and an exception to the preference for grain-fed beef. McDonald's has a strong franchise presence throughout Japan. Hamburgers are seen as a Western food and are especially popular among younger Japanese. In addition to hamburger consumption in restaurants, Japanese also eat a hamburg dish or a hamburg steak at home (a grilled ground beef patty, similar to a hamburger without the bun). These beef products are produced from domestic culled dairy cows and imported frozen beef. Grass-fed frozen beef for use in hamburger patties accounts for a significant portion of the beef imported from Australia (see box, "Ground Beef and Hamburgers").

Beef is also in demand from the cosmopolitan restaurant industry. Like most major cities around the globe, ethnic and foreign restaurants are quite common in Tokyo. One can find menus ranging from Chinese to Italian to American-style steakhouse. Restaurants with European or American-style menus have created increased demand for meat, dairy, and bakery products. Particularly in urban areas, establishments demand more beef and meat products as they seek to create authentic dishes and dining experiences for foreign and native Japanese customers.

Offal[2] has a developed market in Japan as well. These cuts, such as tongue, liver, stomach, and intestine, are used in a variety of dishes for both at-home and away-from-home consumption, such as barbecue, hot pot dishes, stews, and soups. Particularly attractive for exporters, these products can be sold at higher prices in Japan than in their respective domestic markets. The United States supplied most of the imported beef offal to Japan prior to 2004. With other major exporters unable to make up for the lost supply of U.S. product, Japanese beef offal consumption has decreased since the BSE-related trade restrictions on U.S. beef.

Ground Beef and Hamburgers

Japan's consumers buy ground meat at stores to take home and cook, often in the form of a hamburg steak. They also consume ground meat in the form of hamburgers sold by fast-food restaurants or in hamburg steaks served at family restaurants.

Since the mid-1990s, hamburger production for fast food has exceeded production for grocery-store ground beef (box fig. 1). Both hamburger and ground beef production fell in 2001-03. BSE was discovered in Japan's cattle in 2000 and in North American cattle in 2003, and food safety concerns may have prompted the reduction in ground meat production and consumption.

Japanese hamburgers are similar in appearance to U.S. hamburgers, but often incorporate features unique to Japan, such as rice buns and teriyaki flavor.

Typically, they use frozen beef trimmings from Australia and New Zealand and beef from Japanese culled cows. Pork, chicken, and fish may also be ground up for manufacture into "hamburgers." For example, in 2007, the Mos Burger chain began selling hamburgers made from both ground pork and beef (Mos Burger).

Hamburger chains in Japan began to grow rapidly in the early 1970s. The largest chains, McDonald's (with over 3,700 outlets), Mos Burger (with over 1,400 outlets), and Lotteria (over 500 outlets), provide menus for breakfast, lunch, and supper and sell a variety of food and beverages (Otsuka et al., 2009).

After 2003, hamburger production rose quickly, regaining the previous peak (in 2000) of over 130,000 tons. At-home use of ground beef, however, has been stable at about 50,000 tons/year.[1]

How Much Imported Beef Is Used for Ground Beef in Japan?

Data on total beef use for the ground beef/hamburger market are not available. In 2007, about 177,000 tons of ground beef and hamburger were produced (Nihon Hamburg and Hamburger Association).[2] The beef portion must be less than this total, because data for hamburgers include some mixed pork and beef, as well as chicken and fish products, and even the weight of buns.

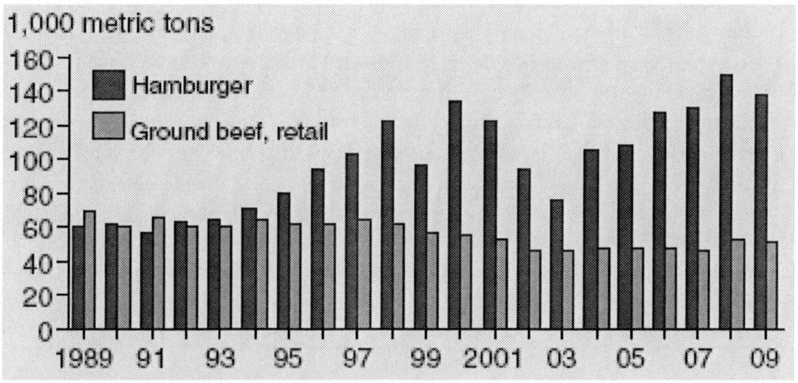

Source: Nihon Hamburg and Hamburger Association.

Notes: Hamburger includes beef, mixtures of beef and pork, chicken, and fish products, as well as buns. Ground meat that is exclusively pork is not included.

Box figure 1. Ground beef and hamburger production in Japan.

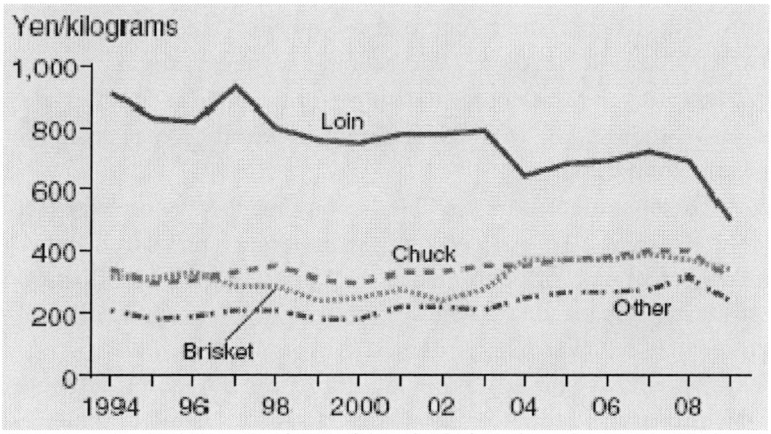

Source: USDA, Economic Research Service calculations based on Japan trade data.

Box figure 2. Unit value of Japan's frozen beef imports.

Trade data provide another view of use of ground meat. Most of the beef imported for hamburgers comes from Australia. Australian data on manufacturing beef, which is often used for hamburger, show exports to Japan of 120,000 tons in 2008. New Zealand data show exports of 17,000 tons of processing beef to Japan—also often used for hamburger. Thus, about 137,000 tons of meat imported by Japan probably was used mostly for ground meat.

The Japanese import category "other meat of bovine animals, frozen" (see fig.

13) is for unspecified boneless cuts and is chiefly imported from Australia and New Zealand. This import category also has the lowest value per unit of the four categories of boneless frozen beef, which suggests that this is low-cost meat from grassfed cattle—the kind of meat widely used around the world for hamburgers (box fig. 2).

However, the "other" category could also contain cuts not accounted for in the remaining three categories (i.e., loins; briskets and plates; chuck, clod, and round) and full sets (the boneless meat from an entire carcass often exported from Australia). After the ban on North American beef in 2003, the quantity and unit value of the "other" category rose, likely because cuts and full sets in this category were replacing some of the meat in the brisket/plate and chuck/ clod/round categories that could no longer be sourced from North America. Before 2003, the quantity of frozen beef imported in the "other" category ranged between 80,000 and 100,000 metric tons, which may have been the upper limit on beef imported for hamburger. The 80,000-ton level represents about 11-12 percent of the pre-2001 beef imports (frozen and chilled) and about 16-17 percent of post-2003 beef imports.

----

[1]Data are not available on ground beef used for hamburg steaks in restaurants.
[2]Excluding all-pork products.

Where Beef Is Purchased

The Japanese retail grocery sector is highly developed and segmented for specific consumer markets:

- Grocery stores range from high-end or specialty markets to more value-oriented stores. All stores, however, provide beef in their meat or butcher sections;
- Large, multi-level mass merchandising stores, such as Ito-Yokado and Jusco, sell meat and produce, as well as clothing, electronics, packaged goods, and other household goods; and
- Higher end department stores sell beef in basement-level grocery sections.

Beef is sold either as thinly sliced cuts or as larger steak-sized cuts. Stores typically display several different cuts of beef, as well as both imported and domestic beef, in small trays and clear plastic. Japanese law requires that all meat have a country-of-origin label, so customers can identify whether beef is domestic or imported. In addition, a national animal identification system allows stores to provide information about domestic beef, some even as detailed as a picture of the farmer or farm that raised the animal. While customers do not always use these traceability features, Japanese retailers insist that their presence assures customers of the safety and integrity of their products (Clemens, 2003).

Japan's numerous convenience stores have become more prominent suppliers in the retail beef market (Matsumoto et al., 2008).[3] Prepared meals are commonly sold in stores, such as 7-Eleven. Bento boxes—prepared lunches with an assortment of meat, fish, rice, and vegetables—are sold in street shops and railway stations, as well as in convenience stores. Prepared meals and prepared meat are popular because of their convenience and availability in larger grocery stores, in addition to the smaller shops, stands, and convenience stores.

Historical Consumption Patterns

Beef consumption in Japan rose dramatically in 1990 after quantitative import restrictions on beef were removed. From 1990 to 1995, beef consumption per person increased annually by an average of 6 percent, including a 12-percent increase in 1993 (MAFF, 2005). A combination of modest economic growth, greater availability of imported beef, and a developing taste for imported grain-fed beef led to strong increases in consumption during this period.

Growth in beef consumption has repeatedly faltered since the mid 1990s (fig. 2). Consumption had fallen 25 percent by 2007 to 5.7 kilograms (kg) per person, compared with its peak of 7.6 kg per person in 2000 (MAFF, 2005). This decline can be attributed to a series of events that influenced demand for beef and beef products.

Weak economic growth was one reason for the decrease in demand. From 1997 to 2002, gross domestic product (GDP) per person increased only 1 percent; in several years, GDP contracted. With income levels stagnant, consumers became more value conscious, preferring processed products and lower priced cuts for barbequing, as opposed to higher grade loins and other table cuts (Obara, 2000). Household beef consumption fell most steeply over this period (fig. 3). Growth in demand from the restaurant and hotel industries kept consumption from falling even more dramatically prior to the 2001 discovery of BSE in the Japanese herd (Obara, 1999). After the BSE outbreak in Japan, beef demand fell across the board, leading to increased inventories, lower prices, and increased consumption of competing proteins, pork in particular.

Demographic changes also affect beef demand in Japan, although, on a year-to-year basis, they do not have a large effect on beef consumption. Looking over a decade or more, however, the decline in population and the aging of the population structure are likely to affect consumption significantly.

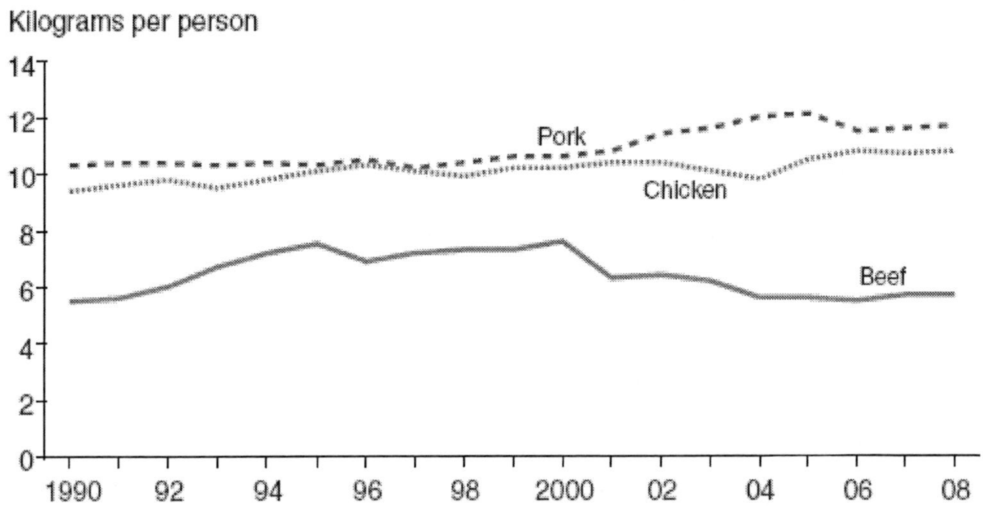

Source: Japan's Ministry of Agriculture, Forestry and Fisheries, Food Balance Sheet.

Figure 2. Beef consumption In Japan has decreased since 2000.

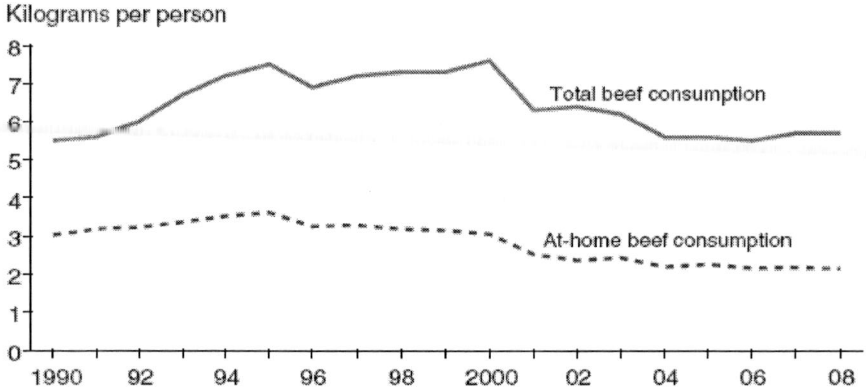

Sources: Japan's Ministry of Agriculture, Forestry and Fisheries, Food Balance Sheet and Statistical Yearbook of the Ministry of Agriculture, Forestry and Fisheries.

Figure 3. At-home beef consumption in Japan has declined since 1995.

Japan's population growth slowed until 2005 when growth reached zero. Since then, population change has been negative. Birth rates in Japan have been very low, as women choose to have fewer children or no children. Japan's population is expected to drop by nearly 2.4 million by 2015 and by over 9 million by 2025, compared with the 2008 level of 127 million (U.S. Census Bureau, 2009). By itself, this decline will reduce consumption of all goods and services, including beef. The population decline has been pronounced throughout rural areas and smaller metropolitan areas, though the larger metropolitan areas of Tokyo and Osaka continue to grow at a very low rate. Consumption will be increasingly concentrated in these urban areas if current population trends continue.

Because of Japan's low birth rate and the increasing lifespan of its citizens, the average age of the population has been increasing. Mori et al. (2006) used an age/period/cohort analysis of urban household expenditure survey data to examine beef consumption per person. Mori et al. found that the cohort eating the largest amount of beef (at home) was born between 1946 and 1950, followed by the group born between 1951 and 1955. All the cohorts born after 1940 had at-home beef consumption above the national average, except the youngest (born between 1976 and 1980). The cohorts born between 1906 and 1940 had below-average consumption. In the coming decade, the number of people in the older (pre-1940) birth cohorts (with below-average beef consumption) will decrease, and Japan's population will be increasingly dominated by the cohorts born after 1940. These post-1940 cohorts represent a solid base for beef consumption. If younger cohorts (born in 1976 or after) eat as much beef as middle-aged cohorts do now, beef consumption per person could rise from current levels. Consumption could fall if younger cohorts continue to eat less beef than their parents.

The 1996 *Escherichia coli* (E. coli) outbreaks in Japan and BSE in Europe increased consumers' awareness of safety issues for beef (Obara, 2000). Even as economic conditions improved at the turn of the century, the 2001 discovery of BSE in the domestic herd and the possibility of infected animals' entering the food supply damaged beef's reputation. Consumer confidence was further eroded by labeling scandals, in which foreign beef was labeled as domestic product (Obara, 2002). These domestic issues had a more severe impact on consumption than did the foreign outbreaks of livestock diseases. After the discovery of BSE in the North American cattle herd in late 2003, Japan banned imports from the United

States and Canada. With no other country able to produce as much grain-fed beef for the Japanese markets as the United States, availability of beef declined, and beef became significantly more expensive. Even though Australia developed its grain-feeding capacity and the United States regained limited access to the Japanese market, consumption of imported beef is lower than it was in the mid-1990s. As a result, supplies of specific cuts remain tight and prices are still well above their 2003 levels (Obara, 2005).The declining availability of beef resulted in consumers' substituting other meats, such as pork and poultry. Pork is especially responsive to changes in beef prices. In 2004, when U.S. beef was no longer available, beef consumption per person fell by 9 percent, while pork consumption per person increased by 7 percent (see fig. 2). Conversely, in the early part of the 1990s, when beef consumption per person increased, pork consumption per person increased only modestly or decreased. Chicken consumption per person also increased 16 percent after 2004. This increase is partially inflated, however, because of lower consumption in 2003 and 2004 due to avian influenza. Chicken consumption increased 5 percent from 2002 to 2007 (MAFF). Chicken has been less of a substitute than pork for beef. Pork and chicken prices remain well below beef, and price-conscious consumers continue to buy those meats at retail grocery stores in high volumes. These purchasing trends have been particularly evident in the large growth of pork imports into Japan (Obara, 2008).

Recent econometric work using Japanese data supports a positive relationship between beef consumption and income. Elasticity estimates, however, do not settle neatly in a narrow range.[4] The lowest recent estimate for a cross-section of households in 1999 is 0.14 and applies only to beef consumed at home (Mori et al., 2006). The highest estimate (also for at-home consumption) comes from a single equation method and is about 1 (Mori et al., 2006). Recent estimates that account for pork and poultry consumption, as well as beef consumption, provide income elasticity estimates of 0.36 and 0.76 for all beef consumption away from and at home (see box, "Econometric Estimation of Elasticities").[5] The income elasticity of total beef consumption appears to be less than 1. A study that disaggregated beef into one imported and two domestic components found significant differences in demand elasticities for the three segments. All had positive income elasticities, however (Mori and Lin, 1994).

Currently, consumer demand for beef is primarily driven by price. As beef prices remain relatively high, beef sales remain limited, while relatively cheaper pork sales have continued to grow. Estimates of own-price elasticities for beef range between -0.4 and -1.32. While econometric estimates indicate that the effects of prices of other animal proteins (pork, poultry, and sometimes fish) on beef consumption are weak, there is evidence that the price of beef strongly impacts pork consumption.

Econometric Estimation of Elasticities

Many studies have measured the elasticities of meat demand in Japan, but most dealt with the time between 1970 and 1990 (Dyck, 1988; Mues et al., 1991). A study by Thompson (2004) measured elasticities over the period 1981-2000. For this study, methods similar to Thompson's were used on data for 1981-2007 (box table 1).[1] Thompson's estimates and the updated estimates used in this report indicate that beef consumption is the most income sensitive of the meats. The estimated beef income elasticity in this study (statistically, weakly significant)[2] is higher than Thompson's result of 0.36.[3] In this study, the own-price elasticity of demand suggests that beef consumption is quite sensitive to the price of beef, with an elasticity of -1.32 (box table 2). A rise in the price of beef also affects pork consumption, as shown by the cross-price elasticity of 0.51 (pork consumption rises 0.51 percent given a 1-percent rise in the beef price). In contrast, a rise in the price of pork (or poultry) has no significant effect on beef consumption.

These results are similar to those observed by Thompson for the earlier period. Relative to Thompson's results, pork and poultry appear less sensitive to changes in the price of beef. Beef demand in Japan is highly segmented, and the price used in the econometric estimation for this study was only for one cut of one type of beef (wagyu); therefore, the results should be used cautiously.[4]

Box table 1. Estimates of income elasticity of demand for beef

Income elasticity	1981-2007 Obara, McConnell, Dyck	1981-2000 Thompson
Beef	0.76	0.36
Poultry	0.02	0.05
Pork	0.38	0.06

Box table 2. Estimates of price elasticity of demand for beef

	1981-2007 (Obara, McConnell, Dyck) Price		
Quantity	Beef	Poultry	Pork
Beef	-1.32	-0.04	-0.01
Poultry	0.29	-0.61	0.91
Pork	0.51	0.33	-0.33
	1981-2000 (Thompson)		
		Price	
Quantity	Beef	Poultry	Pork
Beef	-1.28	0.04	0.28
Poultry	0.60	-0.73	0.01
Pork	0.75	-0.01	-0.91

----

[1]The estimates made for the current study look at relationships over a period that includes the shocks of BSE outbreaks in both Japan and North America. The estimation uses price and quantity data provided by MAFF and the Agriculture and Livestock Industries Corporation (ALIC) from 1981 to 2007 in an Almost Ideal Demand System (AIDS) with meat expenditure treated as an endogenous variable (Thompson, 2004). For further explanation regarding the data and model specifications, see "Appendix 1: Econometric Estimation of Income and Substitution Elasticities."

[2]See Appendix 1. The regression estimate of the parameter relating national income to meat expenditures parameter is statistically different from zero with 88 percent confidence—less than the generally accepted levels of 90 and 95 percent. Hence, we define it as weakly significant.

[3]Income elasticity measures the responsiveness of consumers' purchases of specific goods to changes in income. Spending for an income-elastic good (elasticity greater than 1) is likely to change proportionately more than the change in income, while spending for an income-inelastic good (elasticity less than 1) is likely to change proportionately less than the change in income.

[4]Mori and Lin (1994) probed various segments of the beef market and found large differences in the characteristics of demand for wagyu, dairy, and imported beef.

Domestic Beef Supply

Japanese farms produce a large share of the country's beef supply. The beef herd numbers 2.89 million head, comparable with the number of cattle in Colorado

(USDA/National Agricultural Statistics Service). Beef production relies on significant Government subsidies and on consumers' willingness to pay higher prices for domestic beef. Farmers have generally not received high net returns on cattle operations in recent years and revenues have fallen below costs. A fundamental reason for the relative lack of economic success in raising beef cattle has been Japan's lack of pasture. The land base consists of forested mountains and intensively cropped lowlands. Imported grains and both domestic and imported roughage have replaced the grass that pastures normally provide to cattle. Virtually all of the grain that is fed must be imported across the Pacific, adding to the cost of the feed.

During a critical phase of cattle rearing—breeding and nurturing calves— reliance on feeds in the absence of pasture adds an expense that farmers in much of the rest of the world do not bear. In North America and Oceania— beef exporters to Japan—calves are typically bred and raised on pasture, which is less costly than feeding in confinement. The high price of calves in Japan is a key constraint on the Japanese farmers' ability to supply beef. Calf prices are high because breeding cows have little access to pasture. Cows are confined in and around sheds, and feeds are brought to them. After weaning, calves are usually raised in confinement by farmers who specialize in raising calves, before being sold to feedlots. For Japan, all three aspects of raising beef stock involve feeding, rather than grazing.

Japan's intensive use of feed ingredients, such as corn, includes a transport cost component that is much higher than that paid by U.S. grain-fed beef producers. The feeds include harvested forage, soilage,[6] and concentrated grain-based feeds. About a fifth of the roughage, and virtually all of the grains and meals, is imported. Total transportation costs to Japan (truck, barge, rail, or ocean shipping) are frequently as high as or higher than the cost of corn at the U.S. farmer's delivery point. Additional costs are incurred as corn is trucked from the port in Japan to the feed mill and as compound feed[7] is transported from feed mills to farms.

In recent years, the cost of production for fed cattle has ranged from $3,000 to $8,000 per steer, depending on breed (table 1), compared with the cost in the United States of $1,000 to $1,500 per steer. The largest cost to a Japanese feedlot is the feeder calf (fig. 4 illustrates this for wagyu steers), with recent costs ranging from $900 to $4,400 per calf, depending on breed. Feed, mostly concentrates, is the second most important production cost, at about $2,000 per wagyu feeder calf. In a U.S. feedlot, feed costs are typically much lower.

To build a domestic beef industry in the face of scarce land resources for pasture, Japan's farmers developed a differentiated cattle herd based on the wagyu-breed animals that were used originally as draught animals for agriculture but, from the 1960s on, became solely meat animals (Longworth, 1983).[8] When fed a diet of primarily grains over a relatively long period, beef from wagyu animals becomes tender and highly marbled with intramuscular fat. The meat looks and tastes different from grass-fed beef and from grain-fed beef using other breeds and feeding regimes. In recent years, 60,000- 70,000 cow-calf operations kept about 700,000 wagyu breeding cows.

Japanese consumers have been steady purchasers of relatively stable amounts of wagyu beef. To gain a wider market for larger amounts of domestic beef, however, Japan's farmers need to provide beef at lower prices. The first effort to do this was to use the male calves from the Holstein dairy herd,[9] as well as heifers not kept for milking purposes.[10] Holstein calves have relatively low value when sold as veal calves. Like wagyu calves, they are raised on specialized farms that buy weaned calves and feed them until they are sold to feedlots. As feeder calves, Holstein steers cost feedlots $929 per steer in 2006, compared with $4,365 per wagyu steer. Raising Holsteins compared with wagyu considerably reduces the cost of a fed steer. As a result, the sales price for fed steers is lower for Holstein animals because of the lower cost of calves and feed and because its value to consumers is lower. Wagyu fed steers sold for $8,033 in 2006, compared with $3,283 for a Holstein steer (see table 1).

Japan uses a five-level quality grading system, where five is the highest score. Meat yields (the usable meat on a carcass) range from A (the highest) to C (the lowest). Wagyu carcasses often are given a grade of A-4 or A-5. Carcasses from Holstein fed cattle, however, are usually B-2 or B-3. Thus, Holstein beef is regarded as lower in quality and is priced accordingly in the retail market. Also, yields are lower, reducing the value of the carcass.

In order to improve the value of beef from Holstein calves, Japan's industry developed a cross breed, the F-1 type.[11] Holstein milk cows are artificially inseminated with semen from wagyu bulls.[12] As with wagyu and Holstein calves, the offspring are sold at weaning to specialized calf raisers. Because no F-1 calves are kept for breeding, the number of steers and heifers being raised for feeding are about equal.

Table 1. Comparisons for feedlots in Japan and the United States, 2006

				Feed	Time in	Total	Cattle, at shipment from feedlot		Carcass
	Price	Weight	Age	costs	feedlot	expenses	Price	Weight	price
	Yen/ animal (US$1)	*Kilograms (pounds)*	*Months*	*Yen/ animal (US $1)*	*Months*	*Yen/ animal (US $1)*	*Yen/ animal (US$1)*	*Kilograms (pounds)*	*Yen/kg (US$/lb)*
Wagyu steer	507,593 (4,365)	283.7 (625)	9.4	232,738 (2,001)	19.8	879,078 (7,559)	934,191 (8,033)	716 (1,578)	2,190 (19)
Holstein steer	108,012 (929)	282.3 (622)	7.7	196,135 (1,686)	14.2	366,218 (3,149)	381,826 (3,283)	751.2 (1,656)	969 (8)
F-1 steer2	257,565 (2,215)	266.8 (588)	8.4	240,535 (2,068)	19.2	586,135 (5,040)	604,195 (5,195)	750 (1,653)	1,439 (12)
US steer	$821.05	317.1 (697)	3	$292.28	5.6	4	$1,078.21	568.8 (1,251)	$3.23

Sources: Japan's Ministry of Agriculture, Forestry and Fisheries, Statistical Yearbook of the Ministry of Agriculture, Forestry and Fisheries, 2007-08; and Hoelscher, 2006.

[1] 116.3 Yen/US$, 2006.

[2] F-1 is a common designation for a cross or hybrid between two species or breeds.

[3] There are no comparable surveys provided by Hoelscher or the USDA to indicate average ages of incoming or outgoing cattle in feedlots. 4There are no comparable figures regarding total expenses in comparison to the Japanese published statistics. Differing fixed and labor cost structures among feedlots add complexity when calculating a cost representative of the entire U.S. industry.

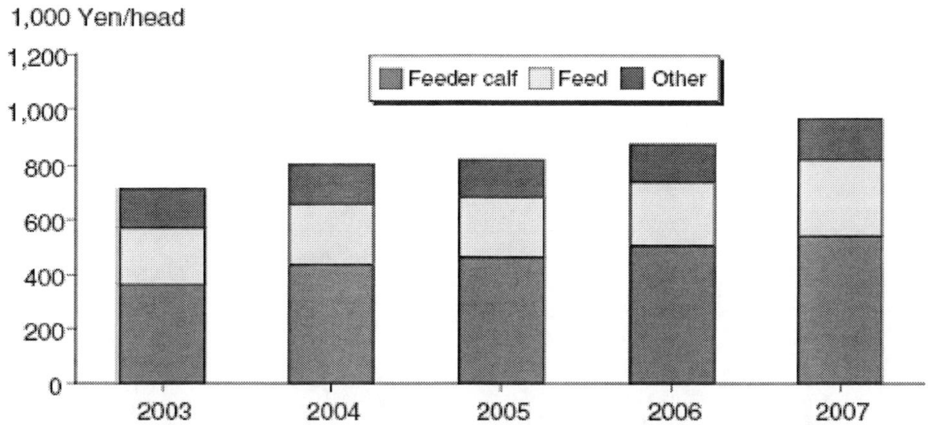

Source: USDA, Economic Research Service calculations based on data from Japan's Ministry of Agriculture, Forestry and Fisheries, Statistical Yearbook of the Ministry of Agriculture, Forestry and Fisheries.

Figure 4. Feedlot costs in Japan, fattening wagyu steers.

Calf prices at feedlot entry are between those for pure Holstein and wagyu calves: F-1 steer prices in 2006 averaged $2,215 (twice that for Holstein steers and just over half the price for wagyu steers). These reduced prices provide significant savings for feedlots. Although F-1 beef most often grades above the Holstein level, grades of B-3 and B-2 still predominate. Despite this, F-1 beef commands a premium over Holstein beef from consumers, and fed F-1 cattle sell for 50 percent or more than Holstein cattle do (MAFF).[13] The relative contribution of different herds to slaughter is shown in figure 5.

Another strategy pursued by the Japanese beef industry is embryo transfer. An embryo is transferred from a wagyu cow to a Holstein cow, producing a wagyu calf. The wagyu cow can then be re-inseminated. Splitting wagyu embryos and then implanting them in two Holstein cows or generating twins in one cow further exploits the wagyu genetic stock. Embryo transfer has proved to be expensive and not always successful. As a result, embryo transfer has not become widespread. In the absence of a technological break-through like embryo transfer or embryo splitting and twinning, the number of calves that Japan can produce is limited. The Holstein milking herd is gradually shrinking. Milk demand is static, and greater milk yields per cow mean that fewer cows are needed each year.

Intensive feeding remains central to the success of Japanese beef production. Wagyu cattle are usually fed longer and go to slaughter at a significantly older age than grain-fed cattle in North America or Oceania. Official estimates show that wagyu cattle average 19.8 months (for 2006) in a feedlot. In contrast, Holstein fed cattle spend 14.2 months in feedlots and are slaughtered at 21.9 months of age—only slightly older than fed cattle slaughtered in the United States. F-1 cattle are fed almost as long as wagyu: 19.2 months (fig. 6). The 5 additional months of feeding for wagyu and F-1 cattle adds to production costs.

Because net returns to farmers increase if fewer or cheaper inputs can be used to produce output of the same value, increasing feed efficiency is often a goal of livestock feeders. A feed conversion rate shows the quantity of feed used to produce a given amount of meat. In 1966, the quantity of feed concentrate fed to cattle in Japan was 1.65 times that of beef production in that year (otherfeeds, not concentrates purchased from feed mills, may have been fed in larger quantities than they are today). In 1986-88, just before the beef import quota began to be phased out, the ratio was 5.28.[14] Some analysts expected the concentrated feed-use ratio to decline, as farmers tried to be more efficient in the face of rising competition from imported beef.

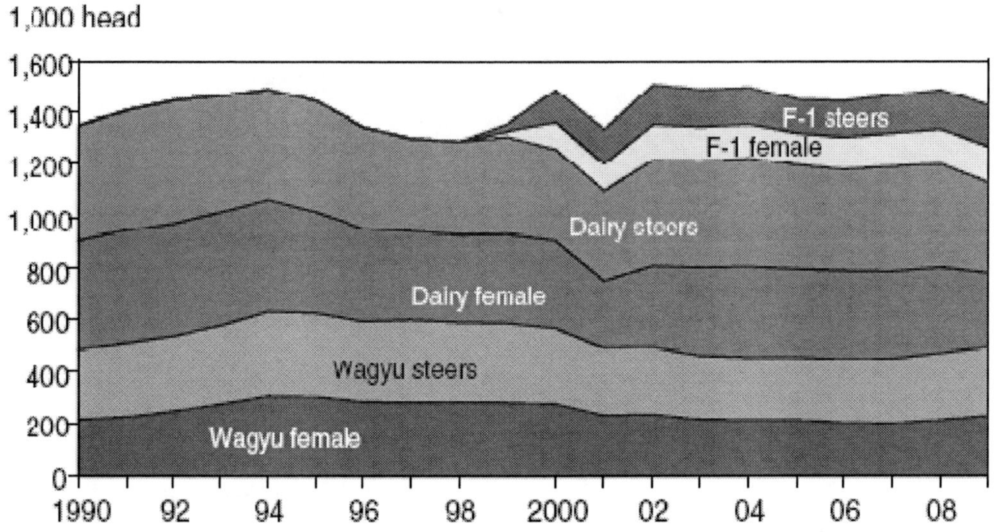

F-1 = Offspring of Holstein cows inseminated with wagyu semen.
Source: Agriculture and Livestock Industries Corporation.

Figure 5. Cattle slaughter in Japan.

The amount of concentrate fed (mostly imported) per unit of meat output has grown, however, reaching a record level in 2008 of 8.96.[15]

The use of compound feed for beef production increased by 27 percent from the average for 1990-92 compared with the average for 2005-07. The additional feed did not significantly increase the amount of meat produced, possibly because Japanese beef fatteners sought to increase the marbling and tenderness of their beef, differentiating it further from imported beef.

Evidently, these beef producers felt the extra cost for increasing the feed-to-meat conversion ratio was offset by greater consumer demand for domestic beef.

Grain dominates the compound feed ration (fig. 7). In recent years, corn has accounted for almost 40 percent of the ration and barley another 16 percent. All the grains, including sorghum, wheat, etc., constitute 61 percent of the compound feed ration. Oilseed meals accounted for 9 percent of the compound feed in 2007, about two-thirds of which was soymeal. Some parts of the feed ration are produced from Japanese-grown materials: rice bran and rice bran meal, wheat bran, beet pulp, and molasses.

Others are products of milling or refining imported basic materials: wheat bran, molasses, and oilseed meal. Most feed ingredients are imported: grains, oilseed meal, alfalfa, corn gluten feed, dairy powders, etc.[16]

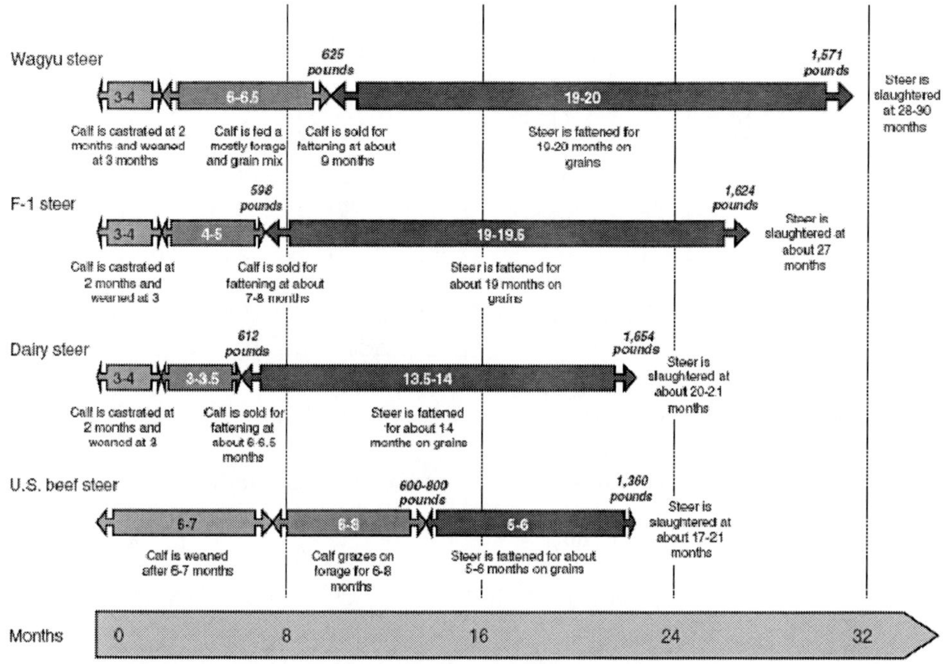

Sources: Japan's Ministry of Agriculture, Forestry and Fisheries, Pocketbook Statistics on Livestock, 2007; Stillman et al, 2009; USDA, National Agricultural Statistics Service, Cattle on Feed and Livestock Slaughter Reports, 2007.

Figure 6. Lifecycle of a steer in Japan.

In recent years, Japan's beef industry has used over 4 million tons per year of compound feed, and feed use for beef has been growing since 2000 (fig. 8). Most of the feed is imported from the United States, especially corn.

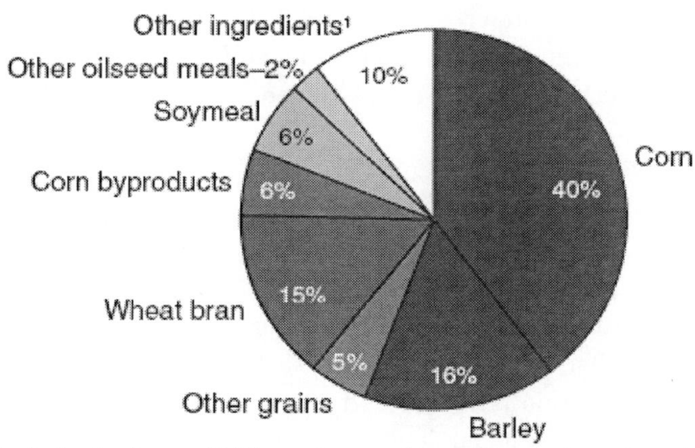

[1] Other ingredients include rice bran, alfalfa, molasses, skim milk powder, whey, beet pulp, and several feed additives.

Sources: USDA, Economic Research Service and Foreign Agricultural Service calculations based on data from the Monthly Feed Statistics of Japan's Feed Supply Stabilization Organization.

Figure 7. Compound feed for beef cattle in Japan, 2007.

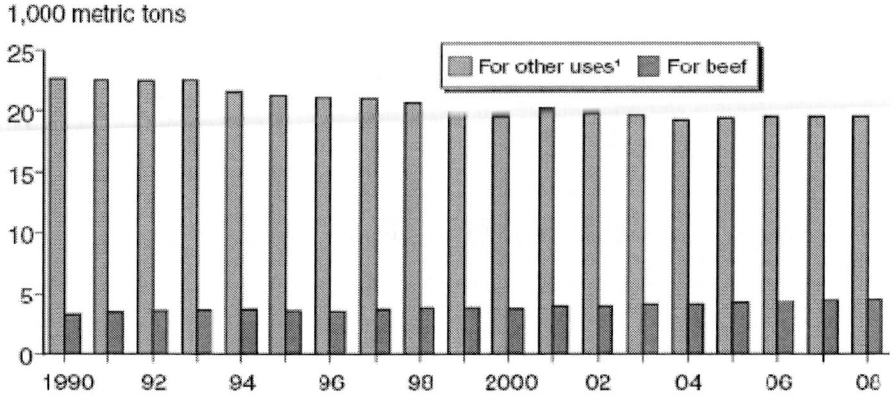

Source: Japan's Ministry of Agriculture, Forestry and Fisheries, Statistical Yearbook of the Ministry of Agriculture, Forestry and Fisheries.

[1] Other uses are for pork, broiler meat, milk, egg, and aquaculture production.

Figure 8. Compound feed production in Japan.

The beef industry in Japan, using 1.8 million tons of U.S. corn, would rank as the eighth largest U.S. corn market, if it were compared with other country destinations for U.S. corn exports. Corn is also a growing market; by 2007, beef fatteners had increased their annual use of U.S. corn by over 300,000 metric tons compared with 1993. Japan's beef industry is the leading user of feed barley and purchases most of the United States' barley exports.[17] In 2008, the value of feedstuffs exported from the United States to supply Japan's beef production was nearly $750 million.[18]

Given the high cost of calves and feed, Japan's domestic beef supply cannot compete with imported beef on price. Two pillars support current supply levels:

- The willingness of Japanese consumers to pay a high price for domestic beef, which can be attributed to a preference for the differentiated, highly marbled beef or a preference for domestic beef over imports, regardless of quality.
- The support for the beef industry by the Government.

Government Policy Affecting the Beef Sector

Japan's Government explicitly targets greater self-sufficiency in agricultural production. Policymakers interpret this to mean that the beef sector, which currently contributes about 6 percent of the value of total agricultural output, should receive Government support and encouragement through domestic programs and a 38.5-percent import duty (MAFF, 2008). Japan is burdened by high Government debt. Government debt, as a percentage of GDP, has been rising since 1992, and exceeded 100 percent in 2009 (International Monetary Fund, 2010). Intense efforts to contain Government spending and reduce the debt burden conflict with a desire to increase support to agriculture, including the beef sector.

Japan's beef sector in 2008 was valued at farmgate at 455 billion yen ($4.4 billion) (Organisation for Economic Cooperation and Development, 2009). Government programs help support the value of beef through direct payments, grants, and loan/insurance subsidies that are financed by taxes and the proceeds of the beef tariff. A full accounting of such support is not available, but a partial sum based on 2008 budgets exceeds 62 billion yen ($600 million). In addition to receiving support from budgetary expenditures, Japanese beef production also receives support from the 38.5-percent import tariff on chilled and frozen beef, which acts as a floor under domestic prices. Japan's Government estimated the value of this market price support at 72 billion yen in 2005 ($662 million) (World Trade Organization, 2009). The Organisation for Economic Cooperation and Development estimated policy-induced transfers to beef producers at 132 billion yen in 2008 ($1.28 billion), based overwhelmingly on market price support from the tariff.[19] Market price support is basically a transfer from consumers to beef producers, when consumers pay higher prices for beef as a result of Government actions that affect prices (i.e., the import tariff). In addition to Government support, the animal feed industry operates a feed price stabilization program (see box, "High Feed Prices in 2008 Shock Japan's Beef Sector").

Major measures taken by the Japanese Government with the 2008 budgetary outlay included the following:[20]

- A deficiency payment for feeder calf operations and other support for raising feeder calves—$236 million;
- A stabilization measure (called MARUKIN) for beef-cattle-fattening operations that provides a safety net when costs exceed incomes— $202 million;
- Payment of more than half the cost of insurance against cattle health risks—$94 million;
- Payment that covers farm and industry BSE-related costs—$98 million;
- Intervention by the government-owned ALIC to stabilize meat prices;
- Loans to farms and loans to prop up the industry feed price stabilization fund—$996 million in subsidized loans; and
- Payment assistance for cattle production and marketing—$130 million.

Support for Calf-Raising Farms: $236 Million Budgeted for JFY2008[21]

A portion of the revenue earned from tariffs on beef imports is channeled to Japan's domestic beef subsidy program—the Deficiency Payment Scheme for Feeder Calves. In JFY2008, 17.6 billion yen ($170 million) was earmarked for deficiency payments. This program is designed to protect domestic feeder calf producers and to secure feeder calf supplies for beef cattle fattening operations. The ALIC,[22] a company owned by MAFF, implements the program and disburses the Government subsidy payments to the feeder calf growers.

Registered feeder calf producers are eligible for deficiency payments when quarterly average auction prices for feeder calves fall below the "guaranteed price" determined by MAFF. The guaranteed price is set for different beef breeds (table 2). The difference between the actual average price and the guaranteed price is paid to the producers (fig. 9), unless prices go below the minimum floor price, or target rationalization price.

When prices fall below the minimum floor price, the Prefectural Feeder Calf Producer Fund provides subsidies. Producers receive 90 percent of the difference between the target rationalization price and quarterly average auction prices of feeder calves in each locality from the Prefectural Fund, in addition to a MAFF deficiency payment that accounts for the difference between the guaranteed price and the target rationalization price.

Feeder calf producers registered in the deficiency payment scheme (table 2) are automatically enrolled in the Prefectural Fund and pay their contribution to the prefectural beef calf producer associations, which operate the fund. These producer contributions constitute a fourth of the fund, prefectural governments contribute another fourth, and MAFF provides funding from its national farm budget through ALIC for the remainder.

High Feed Prices in 2008 Shock Japan's Beef Sector

To protect farmers against price surges, Japan's feed industry has long operated the Feed Price Stabilization Program. Under this program, farmers pay 500 yen per metric ton of concentrated feed into a fund operated by feed millers. Farmers' payments are not subject to Government taxes, such as the income tax. The fund can then be used to reduce feed prices if prices surge. To supplement the farmer-paid fund, an 'umbrella' fund exists with payments supplied 50 percent by the Government and 50 percent by the feed companies. Companies receive payments from the umbrella fund with a 6-month lag (Zennoh, 2008).

Prices of concentrated feeds in Japan soared in 2008, peaking in November 2008 at 73,550 yen per metric ton ($735/metric ton), 50 percent higher than in 2005 and 2006 (MAFF, 2009). This unprecedented increase was caused by higher global grain and oilseed prices, which affected livestock producers worldwide, and the sharp increase in transportation costs due to oil prices. Higher oil prices particularly hurt Japanese livestock producers, who rely upon imported feedstuffs.

Simultaneously, raising the retail price of beef proved impossible. With the economic downturn in 2008, consumers became more cautious about spending, especially for luxury products, such as domestic beef. Japan's producers were squeezed between rising costs and flat or lower output prices.

The Feed Price Stabilization Program had accumulated funds over many years of stable or low world prices for feed ingredients, but the funds were completely used up by April 2008, as agricultural prices surged. Subsidies from the program reduced compound feed prices (for all animal feeds, not only beef feed) by as much as 10,500 yen per metric ton in 2008 (about $97 per metric ton, or 14 percent of the peak price). The Government provided a 35-billion-yen ($338 million) loan to the industry to maintain the fund through JFY2008 (beginning April 1) (Fukuda, 2008).[1] Repaying this loan and rebuilding of the program's funds are expected to be difficult.

----

[1]The Government provided the loan at zero interest, and it must be repaid within 3 years.

During 2000-06, prices paid by feedlots for male dairy calves were below the guaranteed standard price every year and below the target rationalization price in 2003 and 2004 (see table 2).

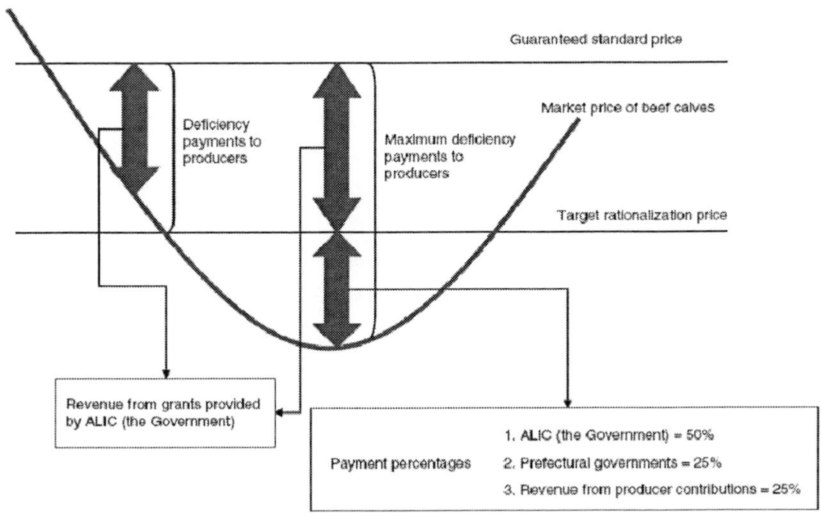

Source: Agriculture and Livestock Industries Corporation.

Figure 9. Payment systems for beef calf producers in Japan.

Table 2. Prices for the deficiency payment scheme for feeder calves in Japan

Item	2000	2001	2002	2003	2004	2005	2006	2007	2008
					Yen per head				
Guaranteed price:									
Black wagyu	304,000	304,000	304,000	304,000	304,000	304,000	304,000	304,000	305,000
Dairy breeds (for beef)	131,000	131,000	131,000	131,000	129,000	110,000	110,000	110,000	113,000
Cross breeds	175,000	175,000	175,000	175,000	175,000	175,000	175,000	NA	178,000
Target rationalization price:									
Dairy breeds (for beef)	80,000	80,000	80,000	80,000	80,000	80,000	80,000	80,000	80,000
Market price:									
Holstein feeder calf	110,710	84,522	100,621	71,674	68,648	81,334	108,012	127,313	NA

Sources: Guaranteed and target rationalization prices are based on USDA, Foreign Agricultural Service GAIN reports JA1002 and JA7058. The Holstein feeder calf price is based on Japan's Ministry of Agriculture, Forestry and Fisheries Statistical Yearbook of the Ministry of Agriculture, Forestry and Fisheries, various issues.

NA = Not available.

In 2004, a farmer selling a calf at the market price of 68,648 yen would have received 49,000 yen (guaranteed price of 129,000 yen less the target rationalization price of 80,000 yen) from the Deficiency Payment Scheme and 10,217 yen (target rationalization price of 80,000 yen less the market price of 68,648 yen, times 0.9) from the Prefectural Feeder Calf

Producer Plan. The 59,217 yen in subsidies would have represented 46 percent of what the farmer received. Clearly, in some years, the Government calf subsidies are critical to maintain the supply of dairy steers for fattening.[23]

The guaranteed standard price for F-1 calves sold to a feedlot is 175,000 yen. During 2003-07, actual market price averages always exceeded the guaranteed price, and no subsidies were awarded. In 2008, however, 4.33 billion yen ($42 million) in subsidies went to F-1 calf raisers, as prices for some F-1 feeder calves fell below the guaranteed standard price (ALIC). Deficiency payments for wagyu calves are uncommon.

In addition to deficiency payments, MAFF administers a subsidy program for cow-calf operators who are willing to increase or maintain the number of breeding cows in their herd during a time of declining calf prices. Payments are made quarterly when the average price per beef calf falls below the trigger levels set per beef breed (i.e., wagyu breeds, such as black wagyu, brown wagyu, etc.) In JFY2008, 6.84 billion yen ($66 million) were allocated to this project.

Support for Beef Farm Operations: $202 Million in JFY2008

In addition to the subsidy for feeder calves, the beef cattle fattening operation stabilization measure (MARUKIN) provides significant support to fattening farms. The Japan Livestock Industry Association administers the program. Farmers must qualify for participation; the Government is trying to limit participation to farmers who specialize in fattening larger numbers of cattle. If labor costs exceed net income from the operation, or if the total cost exceeds gross income, the MARUKIN program covers part of the loss. In 2008, the program provided 20.88 billion yen to farmers ($202 million).

The MARUKIN subsidy is calculated in two ways:

1) If the estimated family labor cost is more than estimated net income, 80 percent of the difference is provided from a fund that is contributed by the Government (75 percent) and farmers (25 percent); or
2) If the operational cost (excluding family labor cost) exceeds gross income, 60 percent of the difference is provided from Government funds.

Hazard Insurance: Over $94 Million in JFY2007

Japan's Government pays half the premiums and much of the administrative costs for the insurance program for beef cattle, as well as supplying reinsurance. Insurance is provided to cover death of farm stock, disease, and injuries. In 2007, the Government payment for premiums was 11 billion yen ($94 million) (National Agriculture Insurance Association, 2009).

The Ministry of Health, Labor, and Welfare (MHLW) paid the costs for BSE testing on all slaughtered cattle through JFY2008, amounting to about 2 billion yen per year (about $19 million).

BSE-Related Measures: $98 Million in JFY2008

Beginning in JFY2009, the MHLW paid for testing only on animals 20 months and older. Prefectural governments have voluntarily paid the costs for testing animals 20 months or younger. Because most of the slaughtered animals are older than 20 months, the amount subsidized is likely to remain near 2 billion yen annually.

In reaction to BSE outbreaks, feed use of meat and bone meal was banned. MAFF spent 8.09 billion yen ($78 million) in JFY2008 to subsidize disposal or alternative uses for the byproducts used in the meal. MAFF also spent 90 million yen (about $900,000) in 2008 on financial support for farms with reported cases of BSE in their cattle.

Beef Price Stabilization

MAFF determines a price band for domestic beef and charges the ALIC with taking market actions when prices fall below or rise above this band (fig. 10).[24] When prices are below the lower bound, ALIC can purchase beef and put it into stocks or ask the industry to stockpile beef with storage costs borne by ALIC.

This process reduces market supply and helps raise prices above the lower bound. When prices are above the upper bound, ALIC can release stocks or purchase beef in foreign markets.

Such actions raise market supply, which puts downward pressure on prices.In several recent years, however, market prices have remained well above the upper bound.

ALIC was evidently unable or unwilling to bring these prices down by releasing additional stocks.

Other Subsidies ($130 Million in JFY2008) and Subsidized Loan Programs

Japan's livestock farmers (both cattle and hog farmers) can tap substantial subsidies for a variety of purposes to contribute to greater efficiency, to satisfy changing environmental regulations, or to retire old debt (table 3). Some measures are multi-year initiatives, while others are for 1 year on an emergency basis. Some measures are designed to assist dairy and hog farms, as well as beef operations.

IMPORTED BEEF

Role in Japan's Market

Over half of the Japanese beef supply comes from imports. Imports primarily come from Australia and the United States, each with specific market niches and constraints. From 1998 until the first cases of North American BSE were discovered in 2003, imports from Australia and the United States accounted for 90-95 percent of total imports. Canada and New Zealand were secondary suppliers. These four countries accounted for about 99 percent of the beef imported into Japan (fig. 11) (Japan Tariff Association, 2008).

Table 3. Selected subsidy and loan programs for beef production in Japan

Project	JFY2008 budget	
	Billion	Million
	yen	US $
Emergency measures to improve beef cattle farm productivity[1]	1.19	12
Comprehensive measures to rationalize meat distribution	2.42	23
Measures to improve demand structure for domestic meat1, [2]	1.3	13
Measures to stabilize demand and supply of livestock byproducts	1.54	15
Individual cattle identification system	0.5	5
Emergency measures to stabilize livestock animal distribution3	0.6	6
Comprehensive measures to use new livestock technologies effectively	0.15	1.45
Create mutual-aid fund for animal disease prevention and control	1.88	18
Maintenance fund to compensate management difficulties caused by livestock disease	0.21	2
Promote effective utilization of livestock animal waste	3	29
Leasing to improve livestock management productivity (loan program)	684	658
Special loan finance for large-size animal farms (cattle)	404	387

Source: Japan's Ministry of Agriculture, Forestry and Fisheries, various web pages.
Note: Exchange rate of 103.4 yen/US$ used.
[1] Also applies to pork.
[2] Also applies to poultry meat. 3Shared with dairy and hog farms. 4Maximum amount of lending.

Japanese and Western consumers generally do not compete for the same beef products on the international market. The demand is strong in Japan for imported cuts of chuck, clod, round, plate, and brisket, which can sell at a premium, compared with the U.S. market (fig. 12). These end meats have more intramuscular fat and are considered more flavorful, which is a preferred characteristic in Japan.

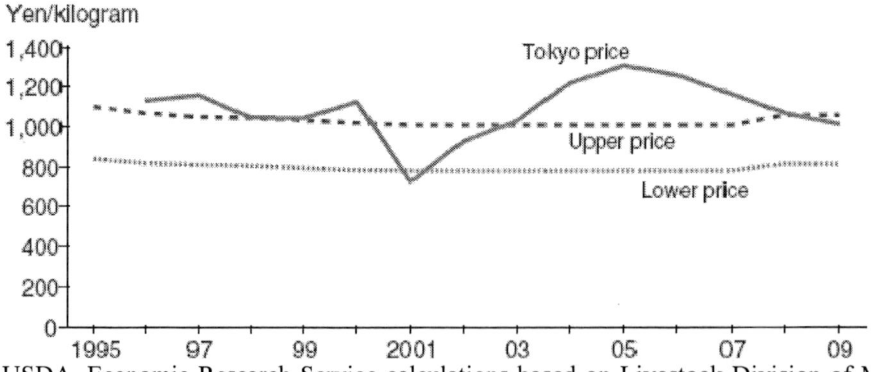

Sources: USDA, Economic Research Service calculations based on Livestock Division of MAFF and Japan Meat Conference joint publication, The Meat Statistics in Japan, and ALIC Monthly Statistics.

Figure 10. Stabilization program for beef, B-2 and B-3 steer carcasses in Japan.

In contrast, Westerners have a stronger preference for middle cuts, such as loin and rib, used for such items as high-value steaks. The Japanese also consume middle cuts for steaks and sliced beef dishes, and prices for middle cuts from wagyu beef are very high. Because of a higher demand from other consumers (e.g., the United States) for imported middle-meat cuts, these cuts are relatively more expensive than imported end cuts, which are perceived as a good value by Japanese buyers. Japanese markets also have a higher demand for offal and organ meats, such as tongue, intestine, liver, and tripe. Consequently, foreign meatpackers rely on Japanese preferences to market otherwise low-valued end cuts and offal, which increases the overall cut-out value and profitability of a carcass (Reed and Saghaian, 2004).

Imported beef is used in a variety of ways. Since imported beef is significantly cheaper than domestic beef (fig. 13), it is eaten more frequently than domestic beef, which is generally saved for special occasions or holidays. Well-marbled domestic beef is reserved for more expensive dishes, such as shabu-shabu, at sukiyaki restaurants and also high-end yakiniku restaurants.

Imported beef is more commonly eaten day-to-day, such as in hamburgers or at popular beef bowl restaurant chains. Imported beef is also available at retail grocery stores for meals at home.

Prior to 2003, most imported beef from Australia and New Zealand was grass-fed. Oceania primarily produced grass-fed beef due to relatively higher grain costs and lower land costs. Australia did export grain-fed beef, which was available at prices competitive with prices for U.S. grain-fed beef. Australian grain-fed cattle were usually on feed for shorter periods than were U.S. grain-fed cattle, which affected the marbling of the beef. Grain-fed beef is more heavily marbled than grass-fed beef, and grain-fed animals produce a white fat generally considered tastier than the yellow fat produced by grass-fed animals. Australia's increased investment in the beef cattle industry and its expanded feedlot capacity has provided more Australian exports of grain-fed beef, although Australia's cattle are still typically not fed for as long as North American cattle are. The United States and Canada export mainly grain-fed beef from cattle that have been fed over 100 days, primarily due to differences in production systems between the United States and Australia.

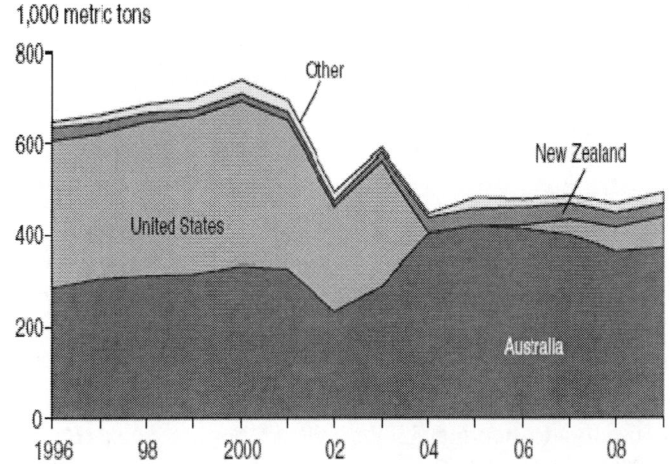

Source: USDA, Economic Research Service calculations based on Japan trade data.

Figure 11. Beef imports in Japan.

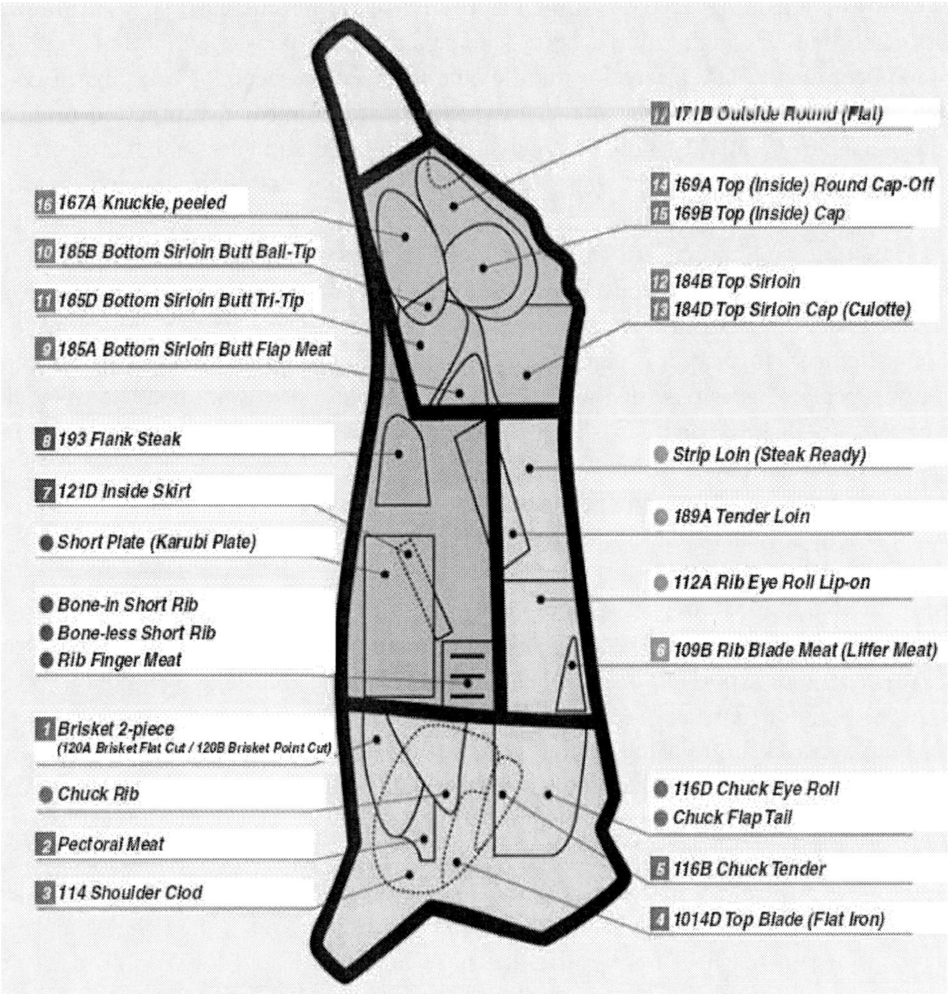

Figure 12. Cuts marketed for the Japanese market.

Because of the relatively high availability of feedgrains, North American production is better suited for finishing cattle with high rations of corn, soybean meal, or other grains.

Geographic factors also influence the type of meat products that the United States and Australia export to Japan. Relatively closer to Asian markets, Australia has an advantage in distance over North America in providing chilled meat cuts (fig. 14). Prior to 2003, about 60 percent of Australian imported muscle cuts were chilled. Conversely, the United States sent mostly frozen product to Japan, with slightly less than 60 percent of American muscle cuts being frozen (fig. 15).

Japanese imports of beef increased steadily until 2000. In 2001 and 2002, the discovery of BSE in the Japanese herd and food safety scandals within Japan hurt overall demand for beef, which caused imports to decline 33 percent from their apex. Beef imports increased again in 2003 before Japan banned U.S. and Canadian imports once BSE had been identified in North America. Imports in 2004 were nearly 40 percent below 2000 levels.

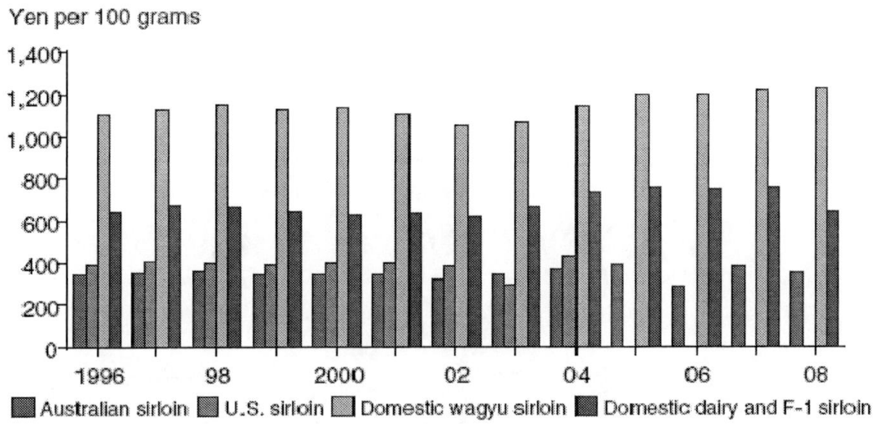

Source: Agriculture and Livestock Industries Corporation.

Figure 13. Retail beef prices in Japan: Imports and domestic.

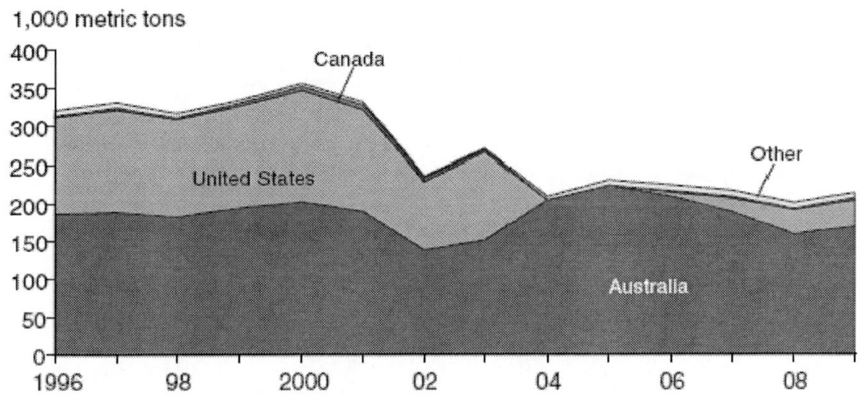

Source: USDA, Economic Research Service calculations based on Japan trade data.

Figure 14. Chilled beef imports to Japan mostly come from Australia.

The ban on U.S. beef in the Japanese market drastically changed the composition of the import market. Prior to the 2003 discovery of BSE, the United States was the second largest foreign supplier of beef behind Australia. Australian and New Zealand producers attempted to fill some of the market void created by the loss of U.S. product. Investment in feedlots and North American-style feeding regimens in both countries tried to capture the grain-fed market segment in Japan.

Total imports, however, never fully recovered and imported beef prices remained above pre-ban levels. Higher feed prices, limited feedlot capacity, and drought hindered the expansion of grain-fed production in Australia. Australian beef was more expensive because of increased demand and often was marketed as full-sets (multiple cuts bundled together in a single order) instead of individual cuts. Thus, Japanese buyers were forced to purchase cuts that were in lower demand, in addition to the cuts that provided better value, which hurt profitability. Trade data show that the sharpest reduction in imports was in the brisket and plate category, where U.S. end meats had dominated (fig. 16). Ultimately, many Japanese consumers reduced their beef consumption, substituting other protein sources, particularly pork, for beef.

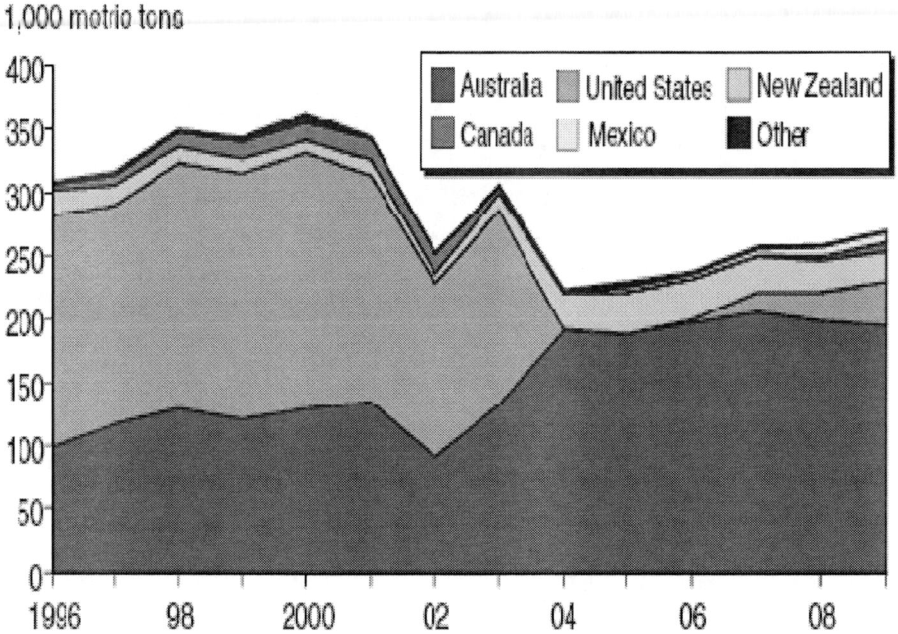

Source: USDA, Economic Research Service calculations based on Japan trade data.

Figure 15. Frozen beef imports in Japan.

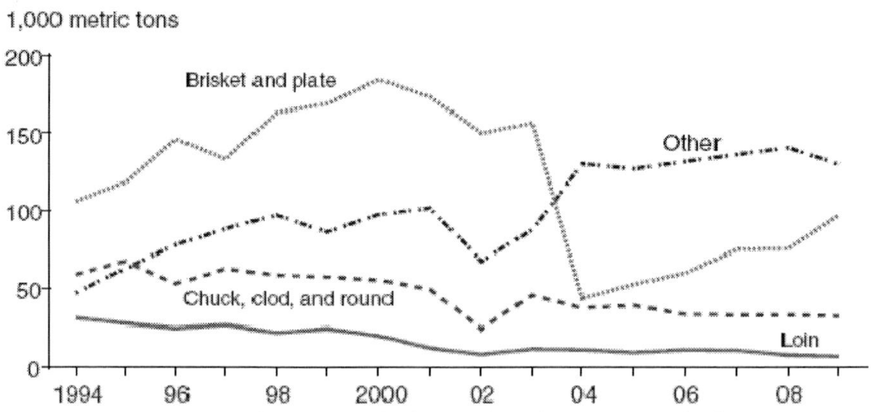

Source: USDA, Economic Research Service calculations based on Japan trade data.

Figure 16. Frozen beef imports by Japan, by cut.

Offal Imports

Japan imports tongues, livers, and other meat that is not considered part of a dressed carcass, such as diaphragm, cheek, or face meat. Before 2004, the United States provided over 75 percent of Japan's sizable imports of bovine offal (fig. 17). During 1994-2000, offal import values ranged from $415 million-$816 million, reaching their peak in 2000. Imports fell precipitously in 2004, after U.S. supply was cut off following the first case of BSE in the United States in 2003. In 2008, the volume imported (35,000 metric tons) was a third of the typical volume imported in 1994-2000 and only 36 percent of the volume imported in 2003, the last year in which U.S. offal was freely imported. As with imported muscle cuts, imported offal from Australia and New Zealand increased after 2004. The volume, however, was not sufficient to fill the void created by the loss of U.S. offal products.

Over the past 5 years, most offal imports have been frozen tongues. From 1995 to 1998, such cuts as cheek and diaphragm meat were the most common offal imports. Diaphragm meat, in particular, was a popular imported cut because of lower tariffs and the ability to substitute it for muscle cuts from a dressed carcass, which carried a higher tariff. Chilled offal cuts have been increasing as a percentage of total offal imports. Although Australia has a geographical advantage in supplying chilled offal, the United States was the primary supplier prior to 2004, and once again became the top supplier in 2008. The United States has yet to recapture its spot as the top supplier of frozen offal.

Reintroduction of U.S. Beef into the Japanese Market

Since U.S. beef was reintroduced into the Japanese market in 2005, total beef imports have increased slightly. U.S. beef has primarily recaptured market share at the expense of Australian beef. The Export Verification (EV) program calls for all U.S. beef to be from cattle younger than 21 months. All products, including offal, must come from plants operating under the EV program, which ensures that all U.S. products exported to Japan are in compliance with the agreement (Clemens, 2007).[25]

Japan's Beef Import Policies

Japan administers two sets of Government policies that directly affect beef imports.

1) The import tariff, or tax, system; and
2) Sanitary rules.

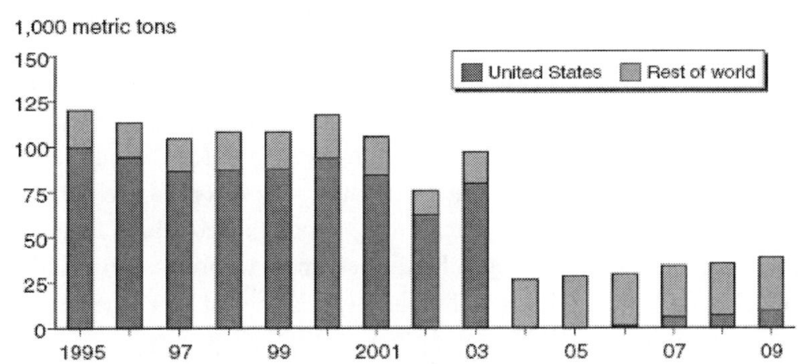

Source: USDA, Economic Research Service calculations based on Japan trade data.

Figure 17. Bovine offal imports by Japan.

Japan's basic import tariff on fresh, chilled, and frozen meat of cattle is 38.5 percent of the value of an import shipment, which applies to all countries, even those holding preferential trading pacts with Japan, and has been in force since 2000 (fig. 18). Two World Trade Organization (WTO) safeguard mechanisms apply to beef import tariffs in the case of rapid rises in import volumes.

- The Uruguay Round Agreement on Safeguards,[26] which Japan has not used for beef.
- A 1995 bilateral agreement between the United States and Japan, which became part of the overall Uruguay Round Agreement, and applies equally to all WTO members exporting beef to Japan.

The bilateral agreement allows Japan to raise the tariff on chilled or frozen beef if imports from the beginning of Japan's fiscal year up to the end of the relevant quarter exceed 117 percent of the quantity of imports during the corresponding period of the preceding year. The increase is then in place until the end of the fiscal year, when it expires. If the situation as described occurs at the end of the fourth quarter of a JFY, Japan can apply a higher tariff during the first quarter of the following JFY. Beef offal is not affected by the safeguard, and the most important offal categories face a 12.8-percent tariff (see "Appendix 2, Japan's Beef Tariffs").

The safeguard gives Japan the option to not raise the tariff, to raise it to a level between 38.5 and 50 percent, or to raise the tariff to 50 percent.

Bound tariffs allow a country to raise its tariff (from whatever level is currently applied) at any time, without foreign consultation. In the case of chilled and frozen beef in Japan, however, the tariff is actually bound at 38.5 percent; only a 17-percent or greater surge in import volume (as described earlier) allows it to apply the 50-percent rate listed as "bound" in international documents.

Raising the tariff to 50 percent when import quantities surge increases the landed cost of a shipment of beef by about 8.3 percent (150/138.5 -1 = .083). This cost increase can represent a serious impediment to trade. The safeguard mechanism also introduces uncertainty into trade. Contracts made when no safeguard is in place can become unprofitable when a safeguard is imposed, but it can be too late to stop the flow of beef and companies suffer losses.

Japan first used the beef safeguard in JFY1996, when it imposed the 50-percent tariff for frozen beef from August 1, 1996, to March 31, 1997 (Obara, 1996). In 2003, Japan applied this safeguard to fresh and chilled beef imports, raising the tariff from 38.5 to 50.0 percent from August 1, 2003, through March 31, 2004 (USDA/ERS, 2009; Obara, 2005).[27]

Tariff increases became more likely after 2004, since the ban on North American beef, imposed because of BSE, resulted in drastic reductions in trade volumes after December 2003. Any major resumption of beef trade in later years would be compared with an artificially low import volume in 2004 and after.[28] Recognizing this, Japan's Government devised a temporary method of choosing a historical import volume with which to compare current import levels. The trigger for a safeguard was set at: 1) 117 percent of the import volume in the corresponding period 1 year earlier; or 2) 117 percent of the average import volume in the corresponding periods in JFY2002 and 2003 (reflecting a period when North American beef imports were quite large) (Obara, 2006).[29] This temporary method, begun in

JFY2006, has been extended annually, preventing imposition of the safeguard through JFY2010.

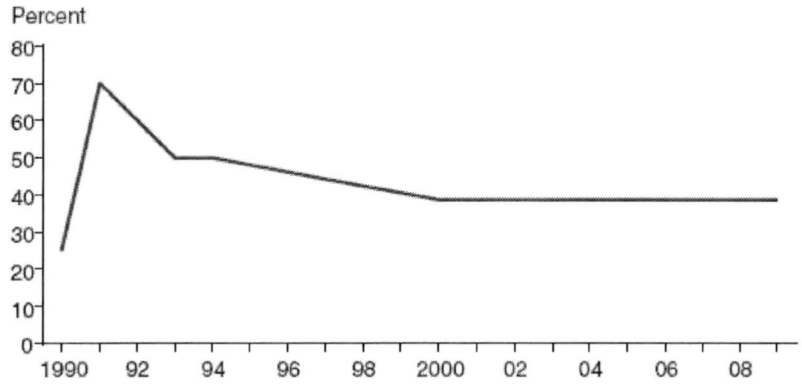

Source: USDA, Economic Research Service.
Notes: A 1988 agreement allowed Japan to raise its tariff after the import quota on beef was phased out, and then required it to be reduced in steps to 50 percent. In the Uruguay Round, Japan agreed to lower the tariff to 38.5 percent.

Figure 18. Japan's tariffs for chilled and frozen beef.

Sanitary rules that have affected beef trade with Japan primarily relate to foot-and-mouth disease (FMD) and bovine spongiform encephalopathy (BSE). Japan refuses to accept uncooked beef from any area not recognized as "free from FMD" or in which cattle have been vaccinated against FMD. For decades, this practice has limited Japan's imports of chilled and frozen beef to supplies from the United States, Canada, Australia, and New Zealand. Minor amounts of chilled and frozen beef have come from a few other suppliers (imports from FMD-free zones in Mexico are the most noteworthy). Imports from South America, a major beef-exporting region, however, have been blocked due to numerous outbreaks of FMD. Japan accepts cooked (thermally treated) beef products from countries known to have FMD.

Japan does not accept beef products, even cooked products, from countries that have uncontrolled risk of BSE outbreaks. In December 2003, Japan ended all imports from the United States because of BSE. Trade was partially resumed 2 years later.

In 2005, Japan reopened beef imports from the United States only for beef from animals 20 months of age or younger at slaughter. Parts designated as Specified Risk Materials (SRMs), such as the brain and spinal column, are excluded from import into Japan. Japan has a similar export verification arrangement with Canada (Canadian Food Inspection Agency, 2008). In practice, the age restriction has been a limiting factor for beef trade with Japan. Age verification entails added costs, and the conservative rules about verifying age from carcass condition likely exclude some animals that are actually 20 months of age or younger. The United States has urged Japan to apply the guidelines recommended by the World Organisation for Animal Health (OIE), which allow trade in beef from carcasses of any age.

Transportation Costs

Most imported beef is brought to Japan by ship. The two most significant ports in Japan are Tokyo and Osaka. American beef is shipped from West Coast ports, particularly from Los

Angeles, Oakland, and Seattle (USDA, Agricultural Marketing Service, 2008). Australian beef is shipped primarily from the ports of Brisbane, Sydney, and Melbourne on the eastern and southern coasts of Australia (Teal et al., 1987).

Shipping from the West Coast of the United States to Tokyo generally takes 10-13 days. According to industry sources, once in Japan, chilled U.S. product generally has a shelf-life of 62 days, compared with 77 days for chilled Australian beef. In 2008, the average container rate was $3,947 for frozen primal beef on a 40-foot-equivalent-unit basis. Rates have remained lower than before the BSE-related trade bans. The container price of frozen primal beef decreased 30 percent from the price in 2003, when the average rate was $4,944 (USDA, Agricultural Marketing Service, 2008).

Shipping costs have increased over the past 3 years. The 2008 financial crisis and dramatic drop in worldwide trade created temporary relief from higher shipping costs, but container rates have steadily increased since U.S. beef was allowed back into the Japanese market. Chilled and frozen beef products must be transported in refrigerated containers to maintain the quality of the product. Because of the limited number of containers for shipping, countries must bid their beef products against poultry and pork exports to Japan, which have greatly increased since 2003. Global competition, especially from Brazil, has also increased the demand and price for shipping containers. Global exports of meat and poultry products increased 32 percent between 2003 and 2008 (USDA, Foreign Agricultural Service, 2009). Companies also must bid for ships to carry their containers. As the volume of worldwide trade increased and the number of ocean vessels remained relatively fixed, container rates also increased, accounting for part of the increase in container rates from 2006 to 2008. As trade volumes and demand for containers declined in the last quarter of 2008, container rates subsequently fell (fig. 19).

Exchange Rates

The strength of the Japanese yen relative to the currencies of major trading partners affects the price of imported beef.[30] The yen has historically been weaker against the U.S. dollar since the late 1990s, compared with the Australian, New Zealand, and Canadian dollars. The exchange rate between the U.S. dollar and the yen, however, is more stable among the major beef exporters (fig. 20).

After reaching a peak in 2000, the yen steadily depreciated against the Canadian, Australian, and New Zealand dollars until the financial crisis in 2008 caused rapid appreciation (University of British Columbia). The exchange rate for the U.S. dollar traded within a much tighter range until the financial crisis in the second half of 2008, when the yen appreciated against it as well. Exchange rates explain some of the fluctuations in imported beef prices, particularly for Australian beef (ALIC).

At the beginning of 2004, when trade bans against U.S. beef had just begun, the yen had become weaker against the Australian dollar, while staying relatively steady against the U.S. dollar (University of British Columbia). This would have made U.S. beef relatively less expensive than Australian beef, all things being equal. BSE-based bans, however, prevented U.S. beef from benefiting from the expected price advantage. Conversely, while trade bans resulted in limited supplies and increased prices in Japan's beef market, the appreciating

Australian dollar further compounded the price increase by making available Australian beef more expensive in terms of yen.

When trade of U.S. beef resumed in 2006 under the EV Program, the yen continued to appreciate against the U.S. dollar and depreciate against the Australian dollar. This combination meant that U.S. beef products were relatively cheaper in terms of yen. This pattern changed, however, after the summer of 2008's financial crisis. As a result, the yen appreciated rapidly against nearly all major currencies, except for the U.S. dollar, against which it appreciated much more modestly in percentage terms. The pattern of convergence between the U.S. and Australian dollar that had taken place over the previous 6 years was temporarily reversed, making Australian products relatively cheaper than U.S. products (University of British Columbia, 2009). In 2009, the Australian and U.S. dollar moved closer in value, with U.S. beef expected to be more competitive. After initial concerns due to the financial crisis were moderated, the Japanese yen was stronger overall, facilitating more Japanese imports of all goods, including beef.

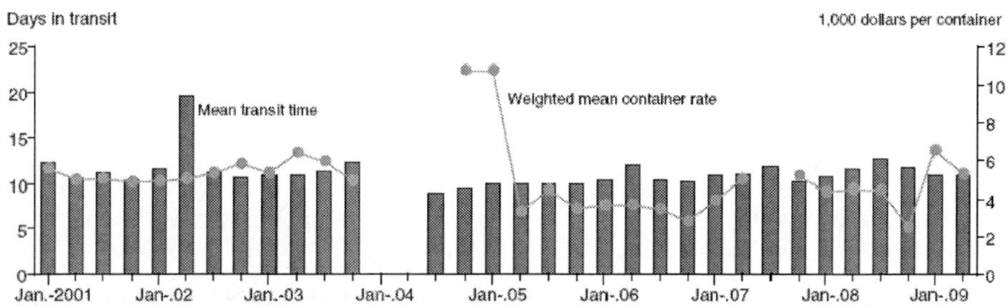

Source: USDA, Agricultural Marketing Service.

Note: Container data not available during period of trade restrictions.

Figure 19. Transportation costs from the United States to Japan.

PRICES

Domestic

Japan's highly segmented beef market is differentiated by price. Even within the same yield and quality grade, carcasses of different breed and sex receive significantly different prices.[31] A top grade wagyu steer has a wholesale price almost twice as high as an F-1 cross-bred steer and nearly three times as high as for a dairy steer. Meat from wagyu cows and heifers carries a premium price over steers in each grade. For dairy and cross-bred breeds, however, meat from heifers is lower priced than that from steers. Breed and place of origin are given on retail labels, while carcass grades are not.

Domestic beef is considerably more costly than imported beef because of the high feed costs and longer feeding periods, but also because consumers are willing to pay prices that cover these costs. Prices for domestic beef dropped in 2001 after discovery of BSE in the Japanese herd and the realization that cattle infected with the disease had entered the food supply. Domestic beef prices gradually recovered in 2002 and 2003, before the discovery of BSE in the U.S. herd in late 2003.

The loss of U.S. beef in the Japanese market created a void in grain-fed beef, particularly affecting lower grades of Japanese beef, which are closer substitutes for U.S. beef. Domestic wholesale beef prices increased 11 percent on average between 2003 and 2007. Grade B beef prices increased more than high-quality wagyu beef. Wagyu beef prices increased during this period but only 5 percent, compared with the 12-percent increase in dairy and crossbred beef prices. Beef from dairy cows increased 24 percent for C-2 and 34 percent for C-1 cows (ALIC).

Retail beef prices rose even more than wholesale prices did. As with wholesale prices, retail prices for nonwagyu beef increased more than retail prices for wagyu beef. Prices for brisket and chuck seemed particularly sensitive; the normal selling price of nonwagyu brisket increased 48 percent between 2003 and 2007, compared with only 13 percent for sirloin (ALIC). In general, the average discount price rose more than the normal selling price, providing fewer savings opportunities for Japanese consumers. These increases are magnified given that overall consumer price levels have declined since 1998 (Euromonitor).

Imports

Prices for imported U.S. beef have been considerably higher in Japan than domestic U.S. beef prices, even before the ban on U.S. beef. From 1996 to 2003, a U.S. sirloin cut cost nearly twice as much in Japan as a choice-grade sirloin steak in the United States, when the price was converted from yen into dollars per pound. Similarly, a U.S. cut of chuck was about 50 percent higher at retail for Japanese customers than for U.S. customers (ALIC and University of British Columbia). The import tariff and transport costs account for some of the increased margins. Consumer preferences, as well as the unique marketing structure in Japan, also raise prices for Japanese consumers.

Australian beef was generally less expensive in Japan than U.S. beef. Australian brisket was particularly less expensive at retail, priced about 60 percent cheaper per kilogram than U.S. brisket, on average, from 1996 to 2003 (ALIC). U.S. cuts of chuck roll and sirloin had smaller premiums over Australian products, averaging 15 and 11 percent, respectively (ALIC).

Compared with domestic beef prices, however, both Australian and U.S. imported beef were much less expensive. Australian beef was about 65 percent cheaper than wagyu beef and 45-50 percent cheaper than beef from dairy and cross-bred animals. U.S. end cuts, such as brisket and rounds, were priced 60-65 percent less than wagyu beef and 35-40 percent less than dairy or cross-bred beef from 1996 until 2003. Higher priced sirloin from the United States was about 45-55 percent less expensive than wagyu and about 10 percent cheaper than dairy and cross-bred beef (ALIC).

The price of Australian cuts increased after U.S. beef was banned from the Japanese market. Australian chuck roll was nearly 29 percent higher from 2004 to 2007 compared with the price from 4 years prior. Brisket prices also increased by nearly 47 percent as the supply of end meats declined. Prices for Australian end meats have remained high as the Australian industry worked to expand their grain-feeding feedlot system (ALIC). However, Australian imports were never able to completely fill the void left by the United States. Limited feedlot capacity and limited Australian domestic demand, as well as drought, made supplying the Japanese market with sufficient individual cuts difficult for Australian producers, unlike the larger American industry. As a result, some Australian beef exported to Japan was marketed as full sets. Full sets included high-demand cuts, such as end meats, as well as cuts that were of lower demand in Japan, primarily middle meats. Including these cuts in full sets affected marketing strategies and the profitability of wholesalers.

Beef prices have remained higher than they were before 2004 as U.S. beef regained access to the market. The current EV program requirements increase the cost and availability of imported beef from the United States. Since U.S. beef has returned to the market, U.S. beef prices have been much higher than Japanese customers are accustomed to, partially dampening demand. Overall, total beef supplies have not recovered to their 2003 levels, and beef prices have remained relatively high.

Conclusion

Japan's market for imported beef depends on demand and supply within Japan. Demand is below historic levels. However, evidence indicates that the Japanese consume more beef when their incomes rise and when beef prices fall. On the supply side, farm returns have been weak and have prevented substantial herd expansion, even in the absence of North American beef from the market. If future conditions are favorable for higher beef demand, much of the extra consumption will be supplied by imports.

There is no guarantee that Japanese incomes will grow or that beef prices will fall in the coming years. If we imagine a scenario where incomes, prices, and other factors do not change, and we disregard changes in age cohorts, consumption per person will not change. Because Japan's population is declining, total consumption will drop by almost 50,000 metric tons per year by 2020 (fig. 21). Rising beef prices, or a continuation of the economic slump seen in 2008 and 2009, could exacerbate this underlying decline dictated by Japan's demographic conditions.

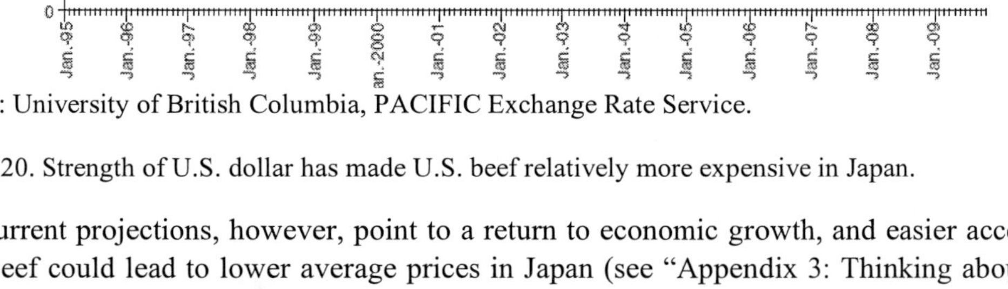

Source: University of British Columbia, PACIFIC Exchange Rate Service.

Figure 20. Strength of U.S. dollar has made U.S. beef relatively more expensive in Japan.

Current projections, however, point to a return to economic growth, and easier access to U.S. beef could lead to lower average prices in Japan (see "Appendix 3: Thinking about the Future of Japan's Beef Market").

For our income projections for Japan, we used those published by the USDA Agricultural Projections to 2019 (USDA/ERS, 2010), and assumed no changes in price or other factors that would affect consumption. We found that beef consumption per person would rise because positive economic growth is projected for 2010 and beyond. The degree to which beef consumption responds to income change is uncertain. An arbitrary choice of a low income elasticity of 0.2 would raise consumption per person, but not by enough to overcome population decline; total consumption would be 13,000 metric tons lower in 2020 than in 2008. Using a higher income elasticity (0.76, estimated for 1981-2007), consumption would rise 86,000 metric tons above current levels by 2020.

The possible effects of changes in beef prices, assuming that all other variables are constant, except for the population decline, can also be examined by making simple assumptions. If we arbitrarily assume that the average beef price declines by 2 percent per year to demonstrate how the elasticity would affect consumption (i.e., more than a 20-percent decline over a 10-year period), and that the own-price elasticity is a low estimate of -0.2 (i.e., for each 1-percent decline in price, consumption per person rises by 0.2 percent), then consumption per person would rise by over 4 percent over a decade. In this case, total consumption would remain near 2008 levels; a population decline would balance price-driven growth in consumption per person. If we assume instead that the own-price elasticity is -1.32 (estimated for 1981- 2007), consumption per person rises much more, and total consumption would be 320,000 metric tons higher in 2020 than in 2008.

Consumption change in these simple scenarios ranges from a 50,000-metricton drop in annual consumption to a 320,000-metric ton rise in 2020, compared with 2008. Much, or even all, of the consumption change would be reflected in import levels. If domestic production were static at 2008 levels, then all the consumption change in 2020 would be reflected in imports (assuming no changes in stocks) that are as much as 50,000 metric tons below 2008 levels or as much as 320,000 metric tons above 2008 levels. If, however, we assume that the proportion of supply coming from domestic production and imports remains the same, then the change is shared between production and trade. Imports could fall by 30,000 metric tons in 2020 or rise by 180,000 metric tons.

Other possible scenarios include a decline in the proportion of supply contributed by domestic production. It is unlikely, however, that only one variable would change. A combination of modestly higher income and modestly lower prices could push domestic consumption above the record high of 1,563,000 metric tons in 2000. Imports could also rise above previous record levels, if domestic production is constant or declines.

If the price of imported beef falls in Japan, import quantities could grow because of the additional demand for beef in response to lower prices and because of potential substitutions away from domestic beef and other meats. Global markets are important in determining imported beef prices, but some price-related factors are specific to Japan's market:

- Imported beef prices could fall if the import tariff on beef were reduced; or
- A change in Japan's policies allows beef from more carcasses to be imported from the United States. The current limitation on beef from animals verified as less than 21 months old is restrictive and could be the cause of higher prices.

While growth in Japan's domestic beef production is not likely and decline is possible, Japan's domestic herds supply a large and distinct portion of beef demand and may continue to do so. Domestic beef production relies heavily on U.S. feedstuffs. Although further intensification of feed use per animal— a trend seen in recent years—is likely to be discouraged by consumers' aversion to foods with a higher fat content (as well as costly for producers to carry out), beef cattle feeding in Japan will remain an important market for U.S. grains, fodder, and oilseed meals.

Japan's beef trade is tied to specific cuts. Restoring former levels will be important to retailers and restaurants that relied on those cuts and want to sell them again.

The conditions prevailing in 2003 and before, where U.S. exporters were able to capture over 50 percent of a larger Japanese import market, may not be fully realized in the future.

The potential Japanese market for U.S. beef, however, could be significantly larger than in 2008.

Appendix 1. Econometric Estimation of Income and Substitution Elasticities

Estimates for the demand elasticities of meats in Japan were calculated using an Almost Ideal Demand System (AIDS), primarily following the method provided by Thompson (2004). What follows is an explanation of the data and methodology used to calculate the estimates reported in the text.

Data

Price and quantity data were used to calculate meat expenditures and a budget share for each good for the AIDS model. The disappearance data series of the Food Balance Sheet (FBS) provided quantities per person for pork, beef, and poultry from 1981 to 2007.

Prices were obtained from the Japan Retail Price Survey series, also from 1981 to 2007. For both beef and pork, shoulder prices were used in the AIDS model. For chicken meat, the broiler leg price was used. Macroeconomic variables were used in an equation relating meat expenditures to total expenditures.

The Japanese Gross Domestic Product (GDP) per person, Consumer Price Index (CPI), and a price index for fresh fish and shellfish were taken from the website of the Ministry of Internal Affairs and Communications, Statistics Bureau.

Model

The model was primarily derived from Thompson (2004), who estimated a nonlinear AIDS model with an added equation that both related meat expenditures to total income and provided an instrument for meat expenditures, which were calculated from prices and quantities, and thus endogenous. Three-stage least squares (IT3SLS in SAS) estimation was used to iterate the added equation and the AIDS budget share equations until parameters converged. Homogeneity in prices and income was imposed on the entire model, and symmetry and adding-up constraints were imposed on the AIDS budget-share equations.

These equations were estimated (variables are in capitals and parameters in lower-case):

Results for the beef and chicken equations were used to calculate the pork parameters and the demand elasticities. Variable definitions and parameter estimates are presented in appendix table 1, and the data used in the regression are found in appendix table 2.

$$Ex = d_o + d_y GDP + d_a P + d_c CPI + d_f P_A + d_l Ex_{t-1} + d_z BSE$$

$$W_B = \alpha_a + \gamma_{aa} P_B + \gamma_{ab} P_C + \beta_a (Ex - P)$$

$$W_C = \alpha_b + \gamma_{ab} P_B + \gamma_{bb} P_C + \beta_b (Ex - P)$$

$$P = \alpha_0 + \alpha_a P_B + \alpha_b P_c + \alpha_c P_P + \frac{1}{2}\left(\gamma_{aa} P_B^2 + \gamma_{ab}(P_B * P_C) + \gamma_{ac}(P_B * P_P) + \right.$$

$$\left. \gamma_{ab}(P_C * P_B) + \gamma_{bb} P_C^2 + \gamma_{bc}(P_C * P_P) + \gamma_{ac}(P_P * P_B) + \gamma_{bc}(P_P * P_C) + \gamma_{cc} P_P^2 \right).$$

where *Ex* is endogenous, and the instruments used to address simultaneity are *CPI, GDP,* Ex_{t-1}, P_B, P_C, *and* P_A.

The meat expenditure term is regressed against GDP, CPI, the price index of fish, meat expenditure lagged 1 year, the calculated meat price index (as specified by the AIDS model), and a dummy variable to capture the effects of BSE in Japan. The dummy variable in the expenditure equation is 1 for all years after 2001, when the first case of BSE was found in Japan. This event, as well as the discovery of BSE in North America in 2003, had a noticeable impact on beef consumption in Japan. The effect is captured in the expenditure equation in this model. The meat expenditure variable is endogenous (Thompson, LaFrance). An instrument for meat expenditure is calculated and then used in the budget share equations, which produce the coefficients used in the elasticity calculations. The parameter estimates are provided below, while the elasticity calculations are included in the main text of the report.

Appendix table 1

Variable	Label
Ex	Log form of total meat expenditure
GDP	Log form of per capita Gross Domestic Product (GDP)
CPI	Log form of Consumer Price Index (CPI)
PA	Log form of fish price index
Ex(t-1)	Log form of total meat expenditure lagged 1 year
BSE	BSE dummy variable
PB	Log form for price of beef
PC	Log form for price of chicken
PP	Log form for price of pork
WB	Budget share for beef
WC	Budget share for chicken
P	Log form of nonlinear price index

Appendix table 2. Model results

	Nonlinear IT3SLS summary of residual errors						
Equation	DF model	DF error	SSE	MSE	Root MSE	R-Square	Adjusted R-Square
WB	2.667	24.33	0.00214	0.000088	0.00938	0.9750	0.9733
WC	2.667	24.33	0.000366	0.000015	0.00388	0.9682	0.9661
Ex	7.667	19.33	0.0322	0.00166	0.0408	0.9726	0.9632

		Nonlinear IT3SLS parameter estimates			
Parameter	Estimate	Approximate standard error	t value	Approximate Pr > |t|	Label
d_y	0.432779	0.2677	1.62	0.1224	
d_a	0.303702	0.2991	1.02	0.3227	
d_c	-0.50671	0.9271	-0.55	0.5910	
d_f	-0.22977	0.7378	-0.31	0.7589	
Y*aa*	-0.16773	0.0200	-8.39	<.0001	
Y*ab*	0.032043	0.0176	1.82	0.0819	
Y*ac*	0.135691	0.0112	12.11	<.0001	
Y*bb*	-0.0055	0.0214	-0.26	0.7992	
Y*bc*	-0.02654	0.0128	-2.08	0.0488	
Y*cc*	-0.10915	0.0122	-8.98	<.0001	
a_a	0.30947	0.0682	4.54	0.0001	
a_b	0.399472	0.0859	4.65	0.0001	
a_c	0.291058	0.0888	3.28	0.0033	
d_0	-1.14257	2.0452	-0.56	0.5829	
d_l	0.882031	0.0969	9.10	<.0001	
d_z	-0.07957	0.0482	-1.65	0.1150	
fi_a	0.417973	0.0236	17.67	<.0001	
fi_b	-0.1389	0.0102	-13.59	<.0001	
Restrict 0	-3.43648	1.3737	-2.50	0.0081	dy + da + dc + df = 0
Restrict 1	-79.8212	47.6526	-1.68	0.0943	g1aa+g1ab +g1ac = 0
Restrict 2	-96.8746	60.3651	-1.60	0.1100	g1ab+g1bb +g1bc = 0
Restrict 3	-71.2183	46.5937	-1.53	0.1291	g1ac+g1bc +g1cc = 0
Restrict 4	-5.03926	2.5778	-1.95	0.0479	aa + ab + ac = 1

Appendix table 3. Data set

	QB	QP	QC	PB	PP	PC	CPI	GDP	PA
1981	3.7	9.6	7.9	5750	2010	1200	79.52987	2648.689	87.9
1982	3.9	9.5	8.3	5710	2060	1180	81.58668	2699.875	93.4
1983	4.2	9.6	8.6	5800	2150	1190	83.15377	2727.916	93.6
1984	4.3	9.7	8.9	5940	2110	1170	85.01469	2816.721	93.6
1985	3.9	9.3	8.4	6230	2100	1150	86.58178	2926.749	95.9
1986	4.2	9.6	9	6360	2040	1120	86.58178	2994.529	97
1987	4.5	10.1	9.3	6380	1990	1060	87.0715	3129.233	95.5
1988	4.9	10.3	9.6	6430	1990	1050	87.7571	3323.787	94.7
1989	5	10.4	9.6	6620	2020	1050	90.20568	3453.953	96.3
1990	5.5	10.3	9.4	6800	2090	1080	93.04603	3647.63	99.8
1991	5.6	10.4	9.6	6900	2150	1090	95.59256	3711.821	103.8
1992	6	10.4	9.8	7050	2190	1120	97.15965	3737.754	105.5
1993	6.7	10.3	9.5	7200	2200	1120	98.43291	3689.357	104.9
1994	7.2	10.4	9.8	7110	2170	1090	98.72674	3759.135	103.1
1995	7.5	10.3	10.1	6890	2190	1080	98.53085	3844.465	101.8
1996	6.9	10.5	10.3	6730	2220	1080	98.92262	3948.099	103.7
1997	7.2	10.2	10.1	7080	2340	1130	100.9794	3938.562	106.1
1998	7.3	10.4	9.9	7280	2370	1150	101.1753	3869.932	107.5
1999	7.3	10.6	10.2	7310	2340	1160	100.6856	3892.479	107.2
2000	7.6	10.6	10.2	7340	2300	1160	100	3983.596	104.9
2001	6.3	10.8	10.4	7690	2300	1170	99.02057	3940.436	104.3
2002	6.4	11.4	10.4	7970	2350	1250	98.43291	3976.587	103.9
2003	6.2	11.6	10.1	8390	2350	1250	98.23702	4054.547	101.9
2004	5.6	12	9.8	8340	2400	1230	98.13908	4132.077	100.6
2005	5.6	12.1	10.5	8440	2390	1230	97.94319	4226.604	100
2006	5.5	11.5	10.6	8550	2390	1230	98.13908	4322.4	102.2
2007	5.665	11.546	10.6636	8840	2420	1250	98.53085	4404.838	103.1

Appendix 2. Japan's Beef Tariffs, 2009

HS tariff lines	Product	Tariff			Note			Trade value, 2008
		WTO		Least- developed countries	Mexico	Chile		
		Ad valorem	specific	ad valorem				
		Percent	Yen/ kilogram	Percent				Million USdollars
0201	Meat of bovine animals, fresh or chilled	38.5		0	30.81	38.5		1,249.687
0202	Meat of bovine animals, frozen	38.5		0	30.82	34.6		895.801

Offal of bovine animals, fresh or chilled:

0206.10.011	Tongues	12.8		0	7.6	12.8		76.087
0206.10.020	Cheek and head meat	50		0	50	50		0.292
0206.10.019	Internal organs	12.8		0	7.6	12.8		75.747
0206.10.090	Other fresh or chilled edible offal	21.3		0	21.3	21.3		0

Offal of bovine animals, frozen:

0206.21.000	Tongues	12.8		0	7.6	11.5		156.296
0206.22.000	Livers	12.8		0	11.5	11.5		2.03
0206.29.020	Cheek and head meat	50		0	30	50		1.886
0206.29.010	Other internal organs	12.8		0	7.6	11.5		43.722
HS tariff lines	Product	Tariff			Note			Trade value, 2008 WTO
		WTO		Least- developed countries	Mexico	Chile		
		Ad valorem	specific	ad valorem				
		Percent	Yen/ kilogram	Percent				
0206.29.090	Other frozen edible offal	21.3		0	19.1	19.1		4.033

Appendix 2. (Continued)

0210.20.000	Meat and edible meat offal of bovine animals, salted, in brine, dried or smoked; edible flours and meals of bovine meat or meat offal		161.5	0	3	3		0
1601.00	Sausages	10		0	10	10		NA

Other prepared or preserved bovine meat, meat offal, or blood:

1602.10	Homogenized preparations	21.3		0		21.3	21.3	19.9 per-cent for Thai-land	0
1602.20.010	Livers	21.3		0		21.3	21.3	17.8 percent for Thailand	NA
	Guts, bladders, and stomachs, whole								
1602.50.100 810, 910	and pieces thereof, simply boiled in water	0		0		0	0		6.314
1602.50.291	Beef or beef offal, simply boiled in water		50		0	50	50		2.836
1602.50.410	Dried after being boiled in water, in airtight containers, not chilled or frozen		38.3		0	38.3	38.3		0
1602.50.420	Dried after being boiled in water, in airtight containers, chilled or frozen		50		0	50	50		0
1602.50.510	Beef jerky, in airtight containers, not chilled or frozen		38.3		0	9	38.3		0
1602.50.520	Beef jerky, in airtight containers, chilled or frozen		50		0	9	50		0
1602.50.590	Beef jerky, other		10		0	9	10		16.051

1602.50.890	In airtight containers, otherwise containing vegetables, not chilled or frozen, not boiled in water	38.3		0	38.3	38.3		0.017
1602.50, 210, 292, 299, 310, 320, 331, 339, 391, 399, 490, 600, 700	Other preparations	21.3		0	21.3[4]	21.3	14.9 percent for Thailand, 1602.50.600	32.258
1602.90.210	Other preparations, including prep-arations of blood of any animal, containing bovine meat or offal	21.3		0	21.3	21.3		NA
1603.00.010	Extracts and juices of bovine meat	12		0	6	6	6 percent for developing countries	NA

Sources: Japan Tariff Association, Customs Tariff Schedules of Japan, 2008. World Trade Atlas trade values.

Note: Least-developed countries include 10 countries in Asia, 33 countries in Sub-Saharan Africa and Haiti.

NA = A value for beef imports cannot be separated because categories report the aggregate of beef- and pork-based imports.

[1] Tariff is 30.8 percent for all boneless cuts; 34.6 percent for bone-in cuts; and 38.5 percent for carcasses;

[2] Tariff is 30.8 percent for bone-in cuts and for boneless loin, chuck, clod, round, brisket, and plate; 34.6 percent for other boneless cuts; and 38.5 percent for carcasses;

[3] Imports are assessed a tariff of 161.5 yen/kg;

[4] Tariff is 19.1 percent for 1602.50.600 and 1602.50.700.

APPENDIX 3. THINKING ABOUT THE FUTURE OF JAPAN'S BEEF MARKET

Reference	*1999*	*2000*	*2001*	*2002*	*2003*	*2004*	*2005*	*2006*	*2007*	*2008*	*2009*	*2010*
						1,000 metric tons						
Historical consumption (1,000 mt) Historical imports (1,000 mt)	1,502 986	1,563 1,045	1,399 982	1,304 697	1,348 833	1,169 634	1,188 686	1,159 678	1182 686	1174 659	1189 672	1195 678
	2011	*2012*	*2013*	*2014*	*2015*	*2016*	*2017*	*2018*	*2019*	*2020*		
						1,000 metric tons						
I. Assessing: Population change, with income and prices fixed												
Consumption, if relative prices, real income, and other effects are constant, but population size changes:												
Case 1: No change in population	1,189	1,189	1,189	1,189	1,189	1,189	1,189	1,189	1,189	1,189		
Case 2: Population change follows U.S. Census projections	1,183	1,180	1,176	1172	1,167	1,162	1,156	1,151	1,144	1,138		
Imports, assuming domestic production fixed:												

Appendix 3. Continued

Reference	*1999*	*2000*	*2001*	*2002*	*2003*	*2004*	*2005*	*2006*	*2007*	*2008*	*2009*	*2010*
						1,000 metric tons						
Case 1: No change in population	664	664	664	664	664	664	664	664	664	664		
Case 2: Population change follows U.S. Census projections	658	655	651	647	642	637	631	626	619	613		
II. Assessing: Income change, with prices fixed, and population declining												
Consumption, if real income changes, while prices remain fixed:												
Case 3: No changes in income	1,183	1,180	1,176	1,172	1,167	1,162	1,156	1,151	1,144	1,138		
Case 4a: USDA baseline projection incomes, elasticity is .2	1,171	1,172	1,173	1,173	1,173	1,173	1,172	1,171	1,169	1,167		
Case 4b: Income elasticity is .76	1,183	1,197	1,211	1,225	1,239	1,252	1,265	1,278	1,290	1,303		
Imports, assuming domestic production fixed:												
Case 3: No changes in income	658	655	651	647	642	637	631	626	619	613		
Case 4a: USDA baseline projec tion incomes, elasticity is .2	646	647	648	648	648	648	647	646	644	642		
Case 4b: USDA baseline projec tion incomes, elasticity is .76	658	672	686	700	714	727	740	753	765	778		
Assessing: Price change, with income fixed and population declining Consumption, if the real price of beef changes, but income does not change:												
Case 5: No beef price change	1,183	1,180	1,176	1,172	1,167	1,162	1,156	1,151	1,144	1,138		
Case 6a: Beef price drops by 2 percent per year, elasticity is -.2	1,176	1,177	1,178	1,178	1,178	1,178	1,177	1,176	1,174	1,172		
Case 6b: Beef price drops by 2 percent per year, elasticity is -1.32	1,229	1,258	1,287	1,316	1,345	1,374	1,404	1,434	1,464	1,494		
Imports, assuming domestic production fixed:												
Case 5: No beef price change	658	655	651	647	642	637	631	626	619	613		
Case 6a: Beef price drops by 2 percent per year, elasticity is -.2	651	652	653	653	653	653	652	651	649	647		
Case 6b: Beef price drops by 2 percent per year, elasticity is -1.32	704	733	762	791	820	849	879	909	939	969		

Source: USDA, Economic Research Service calculations.

ACKNOWLEDGMENTS

The authors gratefully acknowledge the reviews of staff from USDA's Economic Research Service (ERS): William Coyle, Fred Gale, Molly Garber, Kenneth Mathews, Mary Anne Normile, Greg Pompelli, and Richard Stillman; of staff from USDA's Foreign Agricultural Service (FAS): Yong Keun Ban, Hisao Fukuda, and Claire Mezoughem; Shayle Shagam of USDA's World Agricultural Outlook Board; Roxanne Clemens of Iowa State University; Erin Daley of the U.S. Meat Export Federation; John W. Longworth of the University of Queensland; Hiroshi Mori of Senshu University; Michael Reed of the University of Kentucky; James Robb of the Livestock Marketing Information Center; James R. Simpson of the University of Florida and Ryukoku University; and Wyatt Thompson of the University of Missouri.

Many industry experts in Japan generously provided information for this report in briefings and interviews.The assistance provided by Thomas Aoki, Michael Conlon, Midori Iijima, Tamiko Narisawa, Masayuki Otsuka, and Paul Spencer of FAS/Tokyo, and Greg Hanes from Tokyo's U.S. Meat Export Federation is gratefully acknowledged. Linda Atkinson (ERS) provided statistical expertise. Special thanks are extended to Angela Anderson for editorial assistance and Wynnice Pointer-Napper for layout design.

Use of commercial and trade names does not imply approval or constitute endorsement by USDA or ERS.

REFERENCES

Agriculture and Livestock Industries Corporation (ALIC). *Monthly Statistics*, various issues.

Canadian Food Inspection Agency. "Additional Requirements for Establishments Eligible to Export Beef Products to Japan," *Meat and Poultry Manual of Procedures*, annex F to chapter 11, 2008, http://www.inspection.gc.ca/english/fssa/meavia/man/ ch11/coupay/ jap/annexfe.shtml.

Clemens, Roxanne. *Meat Traceability and Consumer Assurance in Japan*, MATRIC Briefing Paper 03-MBP 5, September 2003, http://www.card. iastate.edu/publications/DBS/ PDFFiles/03mbp5.pdf.

Clemens, Roxanne. *After the Ban: The Japanese Market for U.S. Beef*, MATRIC Briefing Paper 07-MBP 12, April 2007, http://www.card.iastate. edu/publications/DBS/ PDFFiles/07mbp12.pdf.

Dyck, John H. *Demand for Meats in Japan; A Review and an Update of Elasticity Estimates*, Staff Report No. AGES880525, U.S. Department of Agriculture, Economic Research Service, August 1988.

Euromonitor International. 2009, http://www.euromonitor.com/.

Fukuda, Hisao, *Annual Japan Grain and Feed Reports*, Global Agricultural Information Network series, U.S. Department of Agriculture, Foreign Agricultural Service (FAS), various issues, http://gain.fas.usda.gov/Pages/ Default.aspx.

Hoelscher, M.A. "Feedyard Analysis," *Feedstuffs,* various issues, 2006.

International Monetary Fund (IMF). W*orld Economic Outlook database*, 2010, http://www.imf.org/external/pubs/ft/weo/2010/01/weodata/index.aspx.

Japan Meat Grading Association. *Beef Carcass Grading Standard*, January 2000.

Japan Tariff Association. *Custom Tariff Schedules of Japan*, 2008.

Japan Tariff Association. "Official trade data of Japan," *World Trade Atlas*, 2009, www.gtis.com

LaFrance, J. T. "When Is Expenditure 'Exogenous' in Separable Demand Models?" *Western Journal of Agricultural Economics* 16(1991): pp. 49–62.

Longworth, John W. *Beef in Japan; Politics, Production, Marketing & Trade*, Australia: University of Queensland Press, 1983.

Matsumoto, Masahiro, and John Mark Looney. *Japan Retail Food Sector Annual Report 2008*, GAIN Series No. JA8713, U.S. Department of Agriculture, Foreign Agricultural Service (FAS), November 14, 2008, http://gain.fas.usda.gov/Pages/Default.aspx.

Miljkovic, Dragan, Gary W. Brester, and John M. Marsh. "Exchange rate pass-through, price discrimination, and U.S. meat export prices," *Applied Economics* Vol. 35, pp. 641-50, 2003.

Ministry of Agriculture, Forestry and Fisheries (MAFF). *Annual Report on Food, Agriculture and Rural Areas in Japan FY 2008 (Summary)*, 2008, http://www.maff.go.jp/e/annual_report/2008/pdf/e_indi.pdf.

Ministry of Agriculture, Forestry and Fisheries (MAFF). *Food Balance Sheet for 2005*, 2007.

Ministry of Agriculture, Forestry and Fisheries (MAFF). *Monthly Statistics of Agriculture, Forestry and Fisheries*, various issues.

Ministry of Agriculture, Forestry and Fisheries (MAFF). *Statistical Yearbook of Ministry of Agriculture, Forestry and Fisheries*, various issues.

Ministry of Agriculture, Forestry and Fisheries (MAFF). *Statistics of Livestock Industry*, various issues.

Ministry of Agriculture, Forestry and Fisheries (MAFF). *Pocketbook Statistics on Livestock, 2007*, July 2008.

Ministry of Agriculture, Forestry and Fisheries, Livestock Division, and Japan Meat Conference. *The Meat Statistics in Japan*, various issues.

Mori, Hiroshi, and Biing-Hwan Lin. *Japanese Beef Market; Distinctly Unique*, Japan: Senshu University Press, 1994.

Mori, Hiroshi, Dennis L. Clason, John Dyck, and William D. Gorman. "Age in Food Demand Analysis," *Cohort Analysis of Japanese Food Consumption—New and Old Generations*, chapter 10, (Mori, Hiroshi, ed.), Japan: Senshu University Press, 2001.

Mori, Hiroshi, Kimiko Ishibashi, Dennis Clason, and John Dyck. "Age-Free Income Elasticities of Demand for Foods: New Evidence from Japan," *The Annual Bulletin of Social Science* Vol. 40, Senshu University, March 2006.

Mos Burger Website. 2009, http://www.mos.co.jp/english/history/.

Mues, Colin, David Harris, Robyn Horton, and Nashika Baskerville. *Substitution relationships between beef and other meats in Japan*, Technical Paper 91.6, Canberra: Australian Bureau of Agricultural and Resource Economics, 1991.

National Agricultural Insurance Association. *The Outline of Japan's Agricultural Insurance Scheme*, briefing booklet, February 2009.

Nihon Hamburg and Hamburger Association. http://nhha.lin.gr.jp/research/ figures.html.

Obara, Kakuyu. *Japan Livestock Reports*, GAIN series, U.S. Department of Agriculture, Foreign Agricultural Service (FAS), various issues, http:// gain.fas.usda.gov/Pages/Default.aspx.

Organisation of Economic Cooperation and Development. *Producer and Consumer Support Estimates*, OECD Database 1986-2007, 2009, http:// www.oecd.org/document/11/0,3343,en _2649_33797_43352651_1_1_1_1,00. html.

Otsuka, Masayuki, Chika Motomura, and Sarah Newsome. *Japan HRI Food Service Sector Report 2009*, GAIN Series No. JA9507, U.S. Department of Agriculture, Foreign Agricultural Service (FAS), March 9, 2009, http://gain.fas.usda.gov/Pages/Default.aspx.

Reed, Michael R., and Sayed H. Saghaian. "Measuring the Intensity of Competition in the Japanese Beef Market," *Journal of Agricultural and Applied Economics* Vol. 36, No. 1, pp. 113-21, April 2004.

Simpson, J.R., Y. Kojima, R. Kada, A. Miyazaki, and T. Yoshida. *Japan's Beef Industry; Economics and Technology for the Year 2000*, United Kingdom: Center for Agriculture and Biosciences International, 1996.

Stillman, Richard, Mildred Haley, and Kenneth Mathews. "Grain Prices Impact Entire Livestock Production Cycle," *Amber Waves*, March 2009, http://www.ers.usda.gov/AmberWaves/ March09/Features/GrainPrices.htm.

Teal, Francis, Andrew Dickson, Darrell Porter, and Diane Whiteford. *Japanese Beef Policies: Implications for trade, prices, and market shares*, Occasional Paper 102, Canberra: Australian Bureau of Agricultural and Resource Economics, 1987, http://www.abare.gov. au/publications_html/ livestock/ archive/87_japanbeef.pdf.

Terauchi, Eiji. Tokyo Meat Market Co., Ltd., interview, September 4, 2008.

Thompson, Wyatt. "Using Elasticities from an Almost Ideal Demand System? Watch Out for Group Expenditure!" *American Journal of Agricultural Economics* Vol. 86, No. 4, pp. 1108-16, November 2004.

Tokyo Metropolitan Central Wholesale Market. *Market Guide*, 2006.

U.S. Department of Agriculture, Agricultural Marketing Service. *Grain Transportation Report*, May 1, 2008, http://www.ams.usda.gov/AMSv1.0/ams. fetchTemplateData.do?template =TemplateA&navID=AgriculturalTransport ation&leftNav=AgriculturalTransportation&page=ATGrainTransportation Report&description=Grain%20Transportation%20Report%20(GTR).

U.S. Department of Agriculture, Agricultural Marketing Service, Audit, Review, and Compliance Branch. *USDA Export Verification (EV) Program Specified Product Requirements for Beef – Japan*, ARC 1030J Procedure, December 12, 2005, http://www. ams.usda.gov/AMSv1.0/getfil e?dDoc Name=STELDEV3103526.

U.S. Department of Agriculture, Economic Research Service. *Japan Briefing Room*, 2009, http://www.ers.usda.gov/Briefing/Baseline/macro.htm.

U.S. Department of Agriculture, , Economic Research Service. *USDA Long-Term Agricultural Projection Tables*, 2010, http://usda.mannlib.cornell. edu/MannUsda/viewStaticPage.do? url=http://usda.mannlib.cornell.edu/ usda/ers/94005/./2010/index.html.

U.S. Department of Agriculture, Foreign Agricultural Service. *Production, Supply and Disappearance (PS&D) Database* 2009, http://www.fas.usda. gov/psdonline/psdHome. aspx.

U.S. Department of Agriculture, National Agricultural Statistics Service. *Quick Stats*. 2009, http://www.nass.usda.gov/Data_and_Statistics/Quick_ Stats/ index.asp.

U.S. Department of Commerce, U.S. Census Bureau. *International Database*, 2009, http://www.census.gov/ipc/www/idb/.

U.S. Department of Labor, Bureau of Labor Statistics. Average price data, 2009, http://data.bls.gov/PDQ/outside.jsp?survey=ap.

U.S. International Trade Commission. *Global Beef Trade: Effects of Animal Health, Sanitary, Food Safety, and Other Measures on U.S. Beef Exports* Investigation No. 332-488, USITC Publication 4033, September 2008, http://www.usitc.gov/publications/332/pub4033.pdf.

University of British Columbia, Sauder School of Business. *PACIFIC Exchange Rate Service*, 2009, http://fx.sauder.ubc.ca/.

World Trade Organization (WTO). Agreement on Safeguards. http://www. wto.org/english/ docs_e/legal_e/25-safeg.pdf.

Zennoh. Briefing, September 4, 2008.

End Notes

[1] A different source, the Ministry of Agriculture, Forestry and Fisheries (MAFF) Meats for Processing survey, cited by the Agriculture and Livestock Industries Corporation (ALIC) in its Monthly Statistics (December 2009), attributes only 1.5 percent of beef consumption to processing.

[2] Offal are products from the animal that are not skeletal muscle. Offal includes inedible products, such as the hide, but also edible organs and muscles not included on a dressed carcass. Edible offal can also be referred to as variety meat.

[3] In 2007, over 68,000 convenience stores were operating in Japan.

[4] Elasticities are a common way of characterizing the effects of income and prices on the consumption of a good. A food demand elasticity is the percentage change in consumption of a food given a percentage change in income or price.

[5] Thompson (2004) estimated 0.36. Obara, McConnell, and Dyck's estimate of 0.76 (this report) shows weak statistical significance.

[6] Soilage is crops cut while green and brought to the cattle without being ensiled. For example, rice, including the grain head, is sometimes cut before grain has ripened and fed to cattle.

[7] Compound feed is a pre-mixed ration for animal feed. Feed mills take component ingredients, such as corn, soybean meal, or byproducts from other milling processes, and produce feed, taking into account certain nutritional requirements and preferences.

[8] "The four modern Japanese breeds are the result of a substantial infusion of European blood during the Meiji Era, together with a Government-sponsored selection program initiated in 1919" (Longworth).

[9] Holstein cows are used primarily for milking purposes in the United States and other countries. Their genetic characteristics are well suited for milk production, but often not ideal for intensive feeding for beef production.

[10] By using Holsteins for beef and developing a market for calves not needed for milk production, Japan's beef industry significantly added to the profitability of dairy operations.

[11] F-1 is a common designation for a cross or hybrid between two species or breeds.

[12] Artificial insemination is almost universal in Japan's cattle industry.

[13] In 2006, the price for Holsteins was $3,283 versus $5,195 for F-1.

[14] This ratio is affected by both the number of animals fed and the duration of the feeding period. The shrinking wagyu herd and increase of dairy animals being fattened in feedlots would decrease the total number of feeding days, and lower the ratio. This effect has been at least partially offset, however, by the addition of the F-1 cross-bred animal to the herd.

[15] Data on compound feed production for beef came from Japan's Statistical Yearbook of the Ministry of Agriculture, Forestry and Fisheries, various editions. Data on beef production came from USDA's Production, Supply, and Disappearance database from 2008. The ratios are a rough estimate of compound feed conversion into meat in Japan's beef industry. Feed for dairy cows that subsequently bear calves (whether the embryo is Holstein, F-1, or wagyu) is not included in this published series from the Japanese Government.

[16] Data on feed composition are provided by Fukuda (USDA, Foreign Agricultural Service/Tokyo) in the annual Grain and Feed Report from 2002 onward and are derived from the Monthly Feed Statistics of Japan's Feed Supply Stabilization Organization.

[17] For example, in 2008, Japan reported imports of 413,662 metric tons of U.S. barley for feed use, which represented 70 percent of U.S. barley exports.

[18] The value of Japan's imports from the United States, including insurance and freight, is calculated by: (1) obtaining Japan's complete data on compound feeds used for raising cattle for beef, published in Monthy Feed Statistics; (2) determining the share of imports in Japan's supply for each of those feeds; (3) determining the U.S. share of Japan's imports of each feed and the import unit value of imports of each feed from the United States; and (4) multiplying the U.S. share of imports of each feed for beef by the import unit value and summing.

[19] This translates to a producer single-commodity transfer equal to 28.75 percent of the value of beef to producers. Both the total support in yen and the percentage of value have remained relatively stable since 1997.

[20] Information about expenditures is incomplete, and amounts cited are often budget allocations, rather than actual expenditures. USDA Foreign Agricultural Service data were coll21JFY means Japan Fiscal Year, which begins April 1 and ends March 31. Thus, JFY2008 is April 1, 2008-March 31, 2009.

[22] Formerly called the Livestock Industry Promotion Corporation (LIPC).

[23] The calf deficiency payment also ensures that there is a market for dairy steers and heifers not put into milk production that can still provide value for the producer, boosting dairy farm income.

[24] This is authorized by the Law Concerning the Stabilization of Livestock Prices.

[25] The United States International Trade Commission, 2008, also provides further discussion on import restrictions and estimates on loss of trade due to BSE-related measures. Details on the export verification program for Japan, administered by USDA's Agricultural Marketing Service can be found at http://www.ams.usda.gov/AMSv1.0/get file?dDocName= STELDEV3103526.

[26] The Agreement on Safeguards is available at the WTO website: http://www.wto.org/english/ docs_e/ legal_e/25-safeg.pdf.

[27] In JFY1996, the tariff on frozen beef imports was 46.2 percent. The tariff was reduced in increments until it reached 38.5 percent on April 1, 2000. Thus, the impact of raising the tariff to 50 percent was greater in 2003 than it was in 1996. See http://www.ers.usda. gov/Briefing/Japan/issuesandanalysis. htm#safeguard for more information.

[28] The JFY2005 first-quarter beef safeguard (April–June) was avoided because meat importers voluntarily delayed customs clearance of shipments that arrived in June until July. As a result, April–June quarterly imports of chilled beef and frozen beef did not exceed the trigger level.

[29] As in the pre-2006 safeguard regime, trade volumes are assessed quarterly, and the cumulative imports through each quarter are compared with comparable periods in the reference year. Thus, imports are compared for April-June (after the first quarter), April-September (after the second quarter), and so forth.

[30] Miljkovic et al. found that the U.S. export unit value of beef destined for Japan dropped by 0.3 percent for a reduction of 10 percent in the value of the yen versus the U.S. dollar, based on data for 1989-96.

[31] Mori and Lin (1994), chapter 5 and appendix A, provide additional discussion that remains pertinent to current conditions.

In: There's The Beef
Editors: Jonathan L. Ruiz and S. R. Fox

ISBN 978-1-61942-943-7

Chapter 5

TRADE, THE EXPANDING MEXICAN BEEF INDUSTRY, AND FEEDLOT AND STOCKER CATTLE PRODUCTION IN MEXICO*

Derrell S. Peel, Kenneth H. Mathews, Jr. and Rachel J. Johnson

ABSTRACT

Cattle raised for export in Mexico represent, on average, over half of total U.S. cattle imports, as the United States has the comparative advantage in feeding cattle. In turn, Mexico imports the largest quantities of U.S. beef and beef products of all U.S. beef trading partners to satisfy the increasing demand for beef, including higher quality feedlot-finished beef.

The authors describe the production systems and stocker and feedlot production practices commonly employed in Mexico, including regional factors like production parameters, cattle marketing, and sourcing practices. Changes in consumer demand and continued population and economic growth support the potential for increased fed beef demand in Mexico.

As a result of increased fed beef production, feed demand from other livestock industries, competition from crop production for land and water to produce feed and forage, and international competition for inputs, the Mexican beef industry will face tremendous challenges in the coming decades.

Keywords: Cattle, Mexico, beef production, feedlot, stocker

* This is an edited, reformatted and augmented version of The United States Department of Agriculture publication, LDP-M-206-01, a Report from the Economic Research Service, from www.ers.usda.gov, dated August 2011.

Mexican Cattle Production Is Important to the United States

Mexico offers extensive forage resources, and over half of the 196 million hectares in Mexico are used for livestock production. With a total cattle inventory of over 23 million head in 2007, cattle are the biggest users of forage, but 7 million head of sheep, 4 million head of goats, and 2 million head of equines (including horses, mules, and donkeys) also compete for forage resources. Mexico's cow-calf industry encompasses a wide array of production activities that vary according to a diverse set of economic motivations, political and legal environments, and cultural adaptations (Peel, Johnson, and Mathews, 2010). Cattle production enterprises produce meat, milk, breeding stock, animals for export, animals for work, and animals for recreation, including fighting bulls and rodeo cattle, largely from dual-purpose cows. This diversity of enterprises occurs in climatic conditions ranging from tropical to desert.

Substantial numbers of cattle are imported into the United States from Mexico, as the United States has the comparative advantage in feeding cattle. Mexican cattle raised for export represent, on average, over half of total U.S. cattle imports. From 1989 to 2010, these imports averaged over 1 million head annually (fig. 1). Imported cattle from Mexico are steers and heifers intended for stockering and/or feeding in U.S. feedlots. They primarily originate from the northern Mexican States (fig. 2) where cattle are predominantly Continental or British[1] breeds and where there is a greater emphasis placed on breeding and genetic improvement to raise cattle suitable for export.

Most of the Mexican cattle imported into the United States are lightweight, weighing less than 400 pounds. Although the lightweight share of cattle imported from Mexico has decreased since 1989, nearly 60 percent of all cattle imports fell into this weight group over the last decade. These cattle are ideal for pasture or stocker (backgrounding) programs, where an emphasis is placed on animal growth and, through the use of harvested or grazed forages, producers can realize relatively low-cost weight gains.

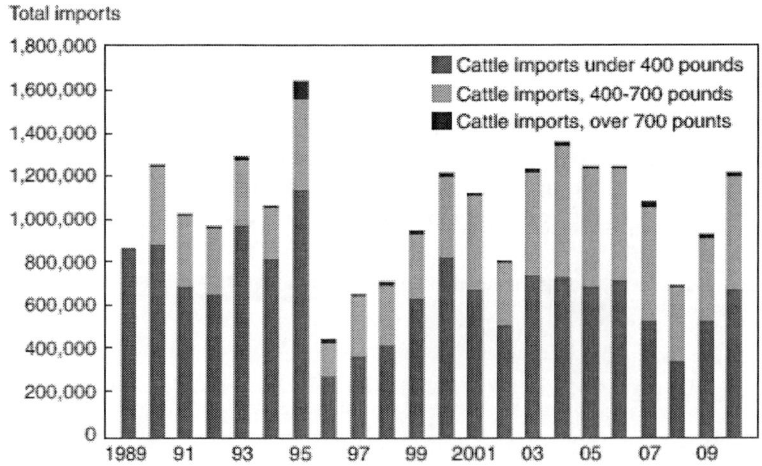

Source: USDA, Economic Research Service.

Figure 1. Total imports of Mexican cattle into the United States, 1989-2010 Total imports.

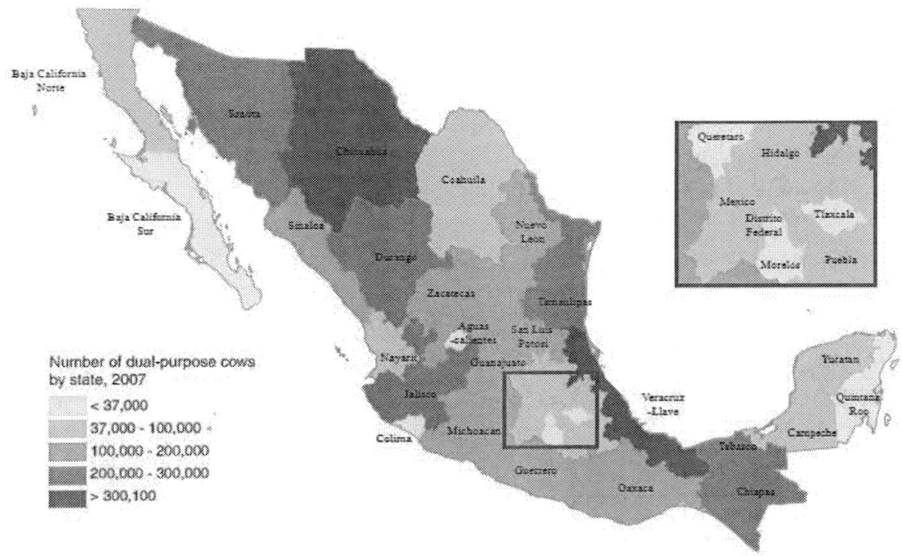

Source: INEGI, Censo Agropecuario, 2007.

Figure 2. Cattle are not uniformly distributed among Mexican States.

Cattle weighing between 400 and 700 pounds may also be stockered or immediately placed in feedlots. Only a small number of cattle imported from Mexico weigh over 700 pounds (see fig. 1), and an insignificant portion of these cattle are for immediate slaughter. Unspayed dairy cattle, especially Holstein and Holstein-crossed females, are prohibited from being imported into the United States from Mexico because of concerns about bovine tuberculosis (TB).

RISING BEEF CONSUMPTION IN MEXICO DRIVES SHIFTS IN MARKET STRUCTURE AND TRADE GROWTH

Despite the effects of the global recession that began in 2008, Mexico's rapidly changing domestic consumer preferences and increasing population have fueled dynamic changes in its beef industry. Increasingly, access to international markets and a growing global beef trade provided some impetus for change, however, population and general economic growth in Mexico prompted the most significant changes in the quantity and quality of meat demanded. Food demand continues to grow rapidly, and meat is gaining an increasing share of the demand for dietary protein.

In 2009, per capita consumption of meats and poultry in Mexico consisted of 39 pounds of beef, 35 pounds of pork, and 65 pounds of broiler meat (USDA, FAS, 2009). Although small compared with U.S. consumption (89 pounds of beef, 64 pounds of pork, and 94 pounds of broiler meat), in the last decade (2000-09), Mexico's per capita consumption grew by more than 10 percent in all major meat categories.

Consumer tastes and preferences for different types of meat are also changing, with demand for fed beef in Mexico growing relative to traditionally produced grass-fed beef.[2] These changes have significant implications for producer selection of feeder cattle production systems and, subsequently, total beef production in Mexico.

As more cattle are raised using semi-intensive and intensive feeder cattle production systems,[3] not only will the number of cattle suitable for export into the U.S. stocker and feeder cattle market increase, but more cattle will likely be slaughtered in Mexico to meet growing consumer demand. Mexico's need for improved market infrastructure and marketing practices, however, is also evident in the expanding Mexican beef sector as beef production competes with grains and alternative meat production systems for resources.

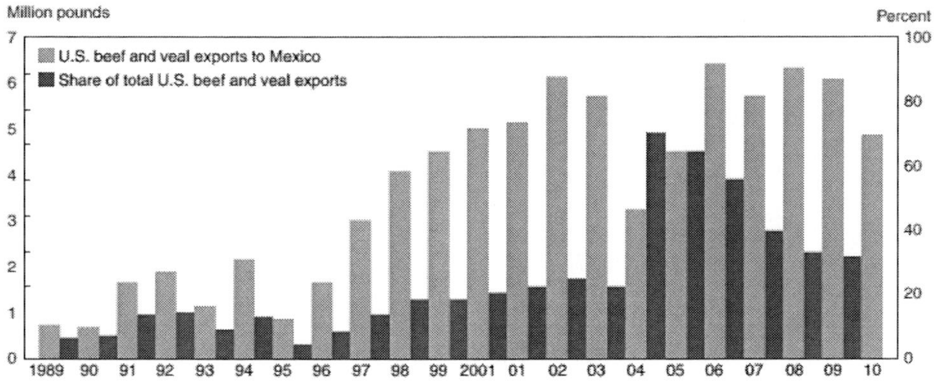

Source: USDA, Economic Research Service.

Figure 3. Exports of U.S. beef and veal to Mexico, 1989-2010.

While Mexican consumers currently prefer a leaner product than U.S. consumers, rapidly growing demand and increased preferences for higher quality meat in Mexico are being met with a larger share of meat exported to Mexico. Mexico is the largest export market for U.S. beef and beef products (fig. 3). Imported meat fills supply gaps not met by domestic production and also provides an improved product mix relative to consumer demands. In addition to the higher prices Mexican producers can obtain for their cattle in the United States, the increasing demand for higher quality meat may partially explain why Mexican cattle are exported to the United States for finishing. Since the United States has the comparative advantage in feeding cattle, due primarily to lower grain costs, Mexico can more effectively export stocker and feeder cattle to the United States and, in turn, import higher quality beef for domestic use.

U.S. beef in Mexico is sold primarily in restaurants, hotels, supermarkets, and warehouse clubs. Since 1997, over 5 million pounds, on average, of U.S. beef is exported annually to Mexico. During 2006-10, beef and veal exports to Mexico accounted for 33 percent of total U.S. beef and veal exports, with this share increasing since the 1990s.

Production of Stocker Cattle in Mexico

Stocker (or backgrounding) programs are defined as grazing or confined production systems that emphasize animal growth and frame development over fat deposition. Stocker cattle production may occur in a wide variety of production systems, ranging from extensive native pasture grazing to intensive confinement. In the traditional Mexican system, where animals are finished with very little fat deposition prior to slaughter, there is little distinction between backgrounding and finishing. Cattle finished in the feedlot generally experience a shorter production period (fig. 4). In Mexico, the most common extensive stocker systems are native pasture grazing, mostly in tropical and temperate areas. More intensive systems include supplementation on pasture, grazing on high-quality irrigated or improved temperate pastures, and confined backgrounding programs.

Until the last decade, stocker production was a mostly nonexistent or at least an unrecognized part of the Mexican beef cattle industry. Historically, calves in the north were exported at weaning, and those not exported joined other cattle as part of the domestic focus on grass-finished cattle. In the traditional pasture finishing system, stocker production is usually an indistinct part of the finishing program. Calves may also be kept for post-weaning development in conjunction with cow-calf operations, but stocker production has typically not been recognized as a separate production enterprise.

As the Mexican beef industry evolves, stocker production will likely become a more distinguishable segment of the industry for two reasons. First, in the north, producers of export calves are increasingly aware of the value of selling heavier cattle. Some calves previously sold at weaning in the fall are now retained on high-quality winter pasture until spring. This process allows for additional weight gain, plus the ability to sell cattle at seasonal price peaks in the spring. The increase in feedlot production also implies an increase in stocker programs. Since the average weaning weight of Mexican calves is roughly 375 pounds (170 kg), it is usually necessary to add 175-265 pounds (80-120 kg) of weight (growth) before cattle are finished in feedlots.

Stocker programs are more than a production process; they also serve an important economic role. Inexpensive forage-based weight gains during the stocker phase are a critical component of the overall efficiency and cost competitiveness of the industry, and these gains are even more critical when relatively expensive grain is used to finish animals. Increased stocker production in Mexico also implies a decrease in export supplies of cattle ideal for the stocker enterprise in the United States. If weights of imported cattle increase, these cattle will be more likely to enter U.S. feedlots directly. Extended stocker programs could also affect total Mexican cattle inventories, although just how is not clear.

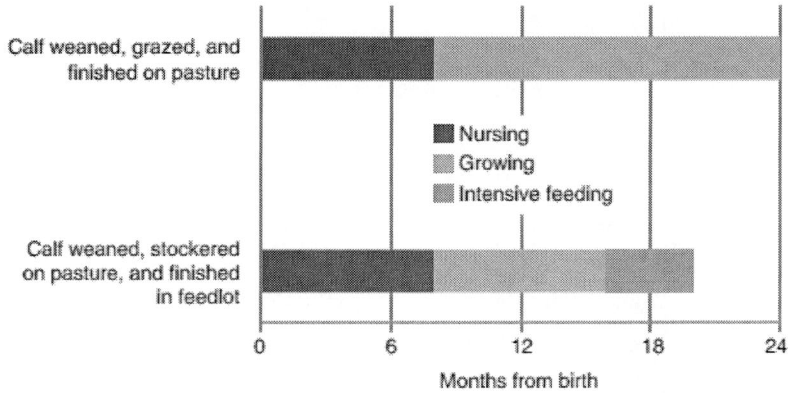

Source: USDA, Economic Research Service.

Figure 4. Production timelines for cattle finished on native pasture versus feedlot.

DYNAMICS AND DEVELOPMENT OF THE MEXICAN FEEDLOT INDUSTRY

Development of the feeding industry in Mexico differs markedly by region. These differences are a function of variations in regional markets, meat preferences, cattle sources, feed availability, competing agricultural activities, and impacts of international trade. In much of Mexico, market preferences for fed beef appear to be increasing rapidly and support increased feeding in areas like the Huasteca and Guadalajara. Reduced cattle supplies and increased competition from the United States, however, have caused some stagnation in the growth of feeding in some traditional feeding areas, including Monterrey, Torreon, and Sonora. In areas like Torreon and Sonora, the development of other agricultural industries, such as dairy, pork, and poultry, creates additional competitive pressures. Producers in these regions have responded by switching to different cattle sources, feed resources, and/or target markets.

Without official data on the Mexican feedlot industry, it is challenging to determine its current status and the rate at which change occurs in the industry. Drought-reduced cattle inventories, increased heifer retention to rebuild cow herds, and rising cattle exports have, at present, reduced feeder cattle supplies. While some areas are increasing feeder cattle numbers, other areas appear to have an excess.

Total fed cattle production depends on feedlot capacity, percent utilization, and the rate of turnover. The rate of turnover depends on the length of the feeding period and the general intensity of feeding operations. Potential turnover rates vary from about 2.5 times per year for feedlots with long feeding periods (130-150 days) to 3 or more times per year for feedlots with shorter feeding periods (90-120 days). The average turnover for the Mexican feeding industry is likely closer to 2.5 times per year. Total productive capability in the industry is not known with certainty, though feedlot production of roughly 2 million head per year is feasible if capacity and turnover rates are fully utilized. The industry appears to be working at 65-75 percent of capacity and, with an annual turnover rate of 2.5 times, is probably producing 1.25 to 1.5 million head per year (table 1).

Table 1. Approximate feedlot capacity, 2011

Region	Head
Mexicali, Baja California	120,000
Sonora	100,000
Comarca Lagunera	60,000
Monterrey, Nuevo Leon	125,000
Tampico/Huasteca	35,000
Culiacan, Sinaloa	90,000
Guadalajara, Jalisco	190,000
Other	60,000
Total	780,000

Source: Author estimates based on industry sources.

The shift from grass-fed beef to grain-fed beef in Mexico has major implications for the overall levels of beef production necessary to meet both changing consumer preferences and increased total demand for meat. Domestic production is limited by forage availability for the cow herd, which is currently dominated by production systems that finish beef on pastures. A general increase in more intensive stocker and finishing systems will release forage resources for increased cow-calf production.

Feeder Cattle Production Systems

Growing consumer demand for meat and rapidly changing consumer preferences for high-quality meat drive the expansion of semi-intensive and intensive cattle production systems in Mexico, particularly in southern Mexican States. Historically, a large percentage of beef in Mexico came from cattle finished on pastures in extensive production systems. Cattle in these extensive systems require 2-3 grazing seasons to reach slaughter weight, and are 3-4 years of age at slaughter. Meat characteristics of grass-fattened animals include a more distinctive meat flavor, a different appearance (yellow fat and darker meat), and a coarser texture (due to older age). A recent survey of seven Mexican packing plants found yellow fat in 13 percent of carcasses (Mendez et al., 2009).

Mexican consumers now prefer lean meat with minimal fat that is white, brighter in color, and has a smoother, more tender texture; such changes in preferences may vary regionally or be more prevalent in urban areas. To achieve the desired results, animals must be intensively fed for a minimal period of time before slaughter to ensure white fat. More intensive production generally results in younger animals at slaughter with brighter meat color and smoother meat texture.

There are two semi-intensive production systems: 1) animals are finished for 70-110 days in confined feeding operations, and 2) animals are finished using high-quality forage and/or supplemental grazing. These systems typically reduce age at slaughter (by 6 to 12 months) to 18-30 months, on average.

In limited instances, cattle may be backgrounded with high quality forage/ supplementation, followed by a feedlot finishing period, or by placing animals in feedlots at lighter weights (and younger age), feeding them longer (130-180 days) and finishing at 13 to 18 months of age. These animals produce very lean carcasses with white fat or, in specific situations, a higher level of fat and marbling. When these animals are slaughtered at the youngest average age, they produce meat of a quality similar to U.S. fed beef.

The implications of increased production intensity reach far beyond the characteristics of the meat. Increased beef production intensity has significant implications for the total level of production possible in the system for two reasons. First, intensive and semi-intensive cattle finishing systems reduce the age at slaughter by at least 1 year, compared with grass-finished cattle, releasing forage for other uses. Second, fed beef production implies at least limited use of concentrate feeds, which are directly substituted for forage. Thus, some fixed forage base, now augmented by use of nonforage feeds, has more total production potential than if a forage-only production system is used. As a result, total beef production potential is increased by using more intensive production systems.

FEEDLOT PRODUCTION PRACTICES IN MEXICO

In Mexico, the majority of animals are placed in feedlots at 528-616 pounds (240-280 kg). In some cases, animals may be placed lighter, as low as 440 pounds (200 kg). Heifers will generally be placed at lighter weights, from 506 to 572 pounds (230-260 kg), while males are generally placed at 550-616 pounds (250-280 kg) but can be placed as high as 660 pounds (300 kg). In a few instances, heifers are placed at up to 700 pounds (350 kg) and bulls at up to 900 pounds (450 kg). In general, cattle are placed heavier and fed for shorter duration in southern areas compared with the north.

Most animals remain on feed for 90-120 days, or occasionally are fed as few as 45 days or as many as 180 days. The length of the feeding period is affected by many factors, including the sex of the animal, the target market, the type of ration fed, and the focus of feedlot management.

Finished cattle weights are generally in the range of 880-990 lbs (400-450 kg). Chilled carcass weights average between 484 and 748 lbs (220-340 kg) (Mendez et al., 2009). Male animals occasionally approach 1,100 lbs (500 kg) and in rare instances up to 1,210 lbs (550 kg). Mexican market preferences are for lean meat with very little external fat. Mendez et al. (2009) reported that 90 percent of carcasses had subcutaneous fat depths of 0.4 inch (1 cm) or less. Finished cattle weights are generally in the range of 880-990 lbs (400-450 kg). Chilled carcass weights average between 484 and 748 lbs (220-340 kg) (Mendez et al., 2009). Male animals occasionally approach 1,100 lbs (500 kg) and in rare instances up to 1,210 lbs (550 kg). Mexican market preferences are for lean meat with very little external fat. Marbling is generally not a strong part of market preference and is not required. In a survey of 7 Mexican packing facilities, Mendez et al. (2009) reported that 90 percent of carcasses had subcutaneous fat depths of 0.4 inch (1 cm) or less, and 94 percent of sampled carcasses had marbling scores of 300 or below 4. Their findings also uncovered distinct differences in consumer preferences in northern markets compared with the remainder of the country, as carcass yield-related traits of northern cattle fattened in feedlots consistently outperformed pasture-finished cattle.

Cattle in feedlots generally gain approximately 2.2-3.3 pounds (1-1.5 kg) per day, with most gains falling between 2.4 and 2.9 pounds (1.1-1.3 kg) per day. In some cases, optimal combinations of genetics, management, and feed quality can result in gains up to 4 pounds (1.8 kg) per day. Feed rations vary widely, and thus, feed conversions similarly vary over a wide range, from 14.3 to 20.9 pounds (6.5-9.5 kg) of feed on a dry-matter basis per pound of gain. Most feedlots average 15.4-17.6 pounds (7-8 kg) on a dry matter basis. Dressing percentages vary from 58 to 63 percent, with most at 59-61 percent.

Rations and Feeding Practices

Feeding management varies considerably across operations. Some operations utilize carefully balanced rations and regular quality testing of feed ingredients, combined with careful feeding and bunk management practices. Other operations use more generalized feed formulations and less precise feed management practices. Feed grains are occasionally rolled or steam flaked, but more often are ground or cracked. Use of variable forage sources, food residuals, and byproduct feeds increases the challenge of providing consistent, balanced rations. Crop production patterns and hot-weather effects cause variations in the seasonal quantity and quality of feed ingredients, increasing the feedlot management challenge.

Most feedlots use a cattle feeding regime that begins with a receiving ration and gradually increases in intensity through three or four stages, while others increase feeding intensity slowly. Feedlot rations can be generalized as principal energy, forage, protein, fat, molasses, and premix.

Principal Energy—Historically, grain sorghum has been the principal feed grain in Mexico. White and yellow corn are both used (and favored) in some regions, while wheat is used predominantly in other areas. Production and use of yellow corn in Mexico has increased dramatically in recent years (Mejia and Peel, 2009). Silage is also used in some instances. In some cases, residual food products, such as bread, tortillas, and snack chips, are important energy sources. If available, industrial byproducts, such as brewer's grains, wheat middlings, rice millings, and fiber residual from sugar cane, are also used. Mexico is the largest importer of dried distillers' grains with solubles (DDGS) from the United States, and exports of DDGS from the United States to Mexico have increased significantly as U.S. ethanol production has grown (fig. 5). Other byproduct feeds include trash and residual from grain cleaning and cotton ginning. These energy feeds typically make up 50- 80 percent of the ration, with the majority in the range of 60-70 percent.

Forage—Forage typically accounts for 8-20 percent of feedlot rations on a dry matter basis. Forage varies from fresh or wilted green chop of ryegrass or other improved grass to relatively high-quality hay from sorghum-sudan or cane to lower quality baled corn or grain sorghum stover. Some of the byproducts discussed above (e.g., sugar cane fiber and gin trash) also provide considerable fiber.

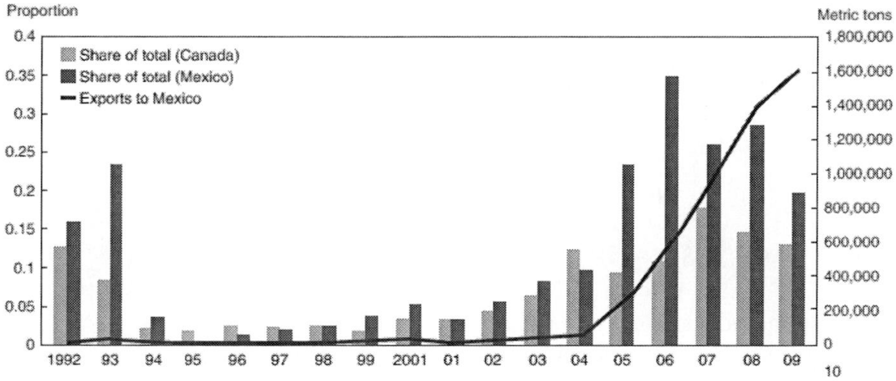

Source: USDA, Economic Research Service.

Figure 5. Exports of U.S. dried distillers' grains with solubles (DDGS) to Mexico and proportions of total DDGS exports to Canada and Mexico, 1992-2009.

Protein—Depending on the energy and forage sources used in the ration, more or less additional protein may be necessary. The most common protein source in many areas is cottonseed meal (or occasionally whole cottonseed). Other common protein sources include soybean meal, fish meal, and poultry meal. Canola and sesame meal are used in some cases. Bioprotein sources, such as yeast-derived protein, are used in a small number of instances.

Production and use of meat and bone meal from ruminants is restricted because of concerns over bovine spongiform encephalitis (BSE). Poultry litter is used to a limited extent, but is consciously avoided in some areas and by some producers.

Fat—Energy in rations is boosted with added fat (3-5 percent) in many cases, primarily from animal fat or a mixture of animal and vegetable fat. Vegetable fat is often preferred due to fear of BSE outbreaks.

Molasses—Molasses is used extensively in Mexican feedlot rations (7-12 percent) because it is widely available due to the relatively large Mexican sugar industry. It controls dust created when feed grains are processed by grinding, and it is an important energy source.

Premix—In most cases, a premix of vitamins, minerals, and ionophores (feed additive that increases feed efficiency and rate of gain) is used in the ration.

Animal Type, Source, and Management

Faced with tremendous variability in cattle quality, market preferences, and interaction with the U.S. market, Mexican feedlots vary widely by region. In the north, where many of the high-quality (No. 1 and 1 1/2) steers are exported, feedlots are composed of half or more No. 1 to 1 1/2 heifers and are supplemented with No. 2 bulls from farther south.[5] Quality of cattle in other feeding areas will vary with equal mixes of No. 2 and lower-quality bulls and heifers to higher proportions of males. In most instances, males are fed as bulls, but some feeding of steers occurs in the north. The majority of irrigated pastures are located in the northwest, allowing for cattle to move to feedlots at a younger age, while feedlot cattle in other parts of the country tend to be somewhat older.

Cattle are sourced from all parts of Mexico, with a large majority of cattle in the northern half of the country either exported or fed in feedlots. Smaller but increasing shares of Gulf Coast and southern cattle are fed in feedlots or finished in semi-intensive supplementation systems on pasture. A limited number of cattle are sourced from Central America,[6] primarily due to disease related issues. Increasingly, cattle are moving from areas farther south in a northward flow to feedlots.

Animal Health and Disease Management

Animal health and disease-related issues restrict movement of cattle between Mexico and the United States, as well as between Mexican States and Central America.

Bovine tuberculosis (TB) is the primary disease that limits Mexico's access to export markets for cattle. TB is endemic in the country, though reasonably well controlled in most areas. Mexico has supported a Federal TB eradication program since 1971 and, in conjunction with individual States, has made substantial progress, particularly in the northern States, where cattle production often is focused on calves for export. TB remains widespread in concentrated areas of dairy production in the northern States, thre atening the status of surrounding beef cattle production areas. In other parts of the country, isolated regions have achieved a TB status that permits cattle exports, but that status is difficult to maintain because of surrounding endemic areas.

Brucellosis is currently endemic and widespread in Mexican cattle. Mexico also has a Federal brucellosis eradication program, but it requires greater resources, and implementation has been slow. The prevalence of brucellosis prevents exports of breeding animals, but its biggest impact is on productivity. Brucellosis causes spontaneous abortion in cattle.

Rabies in cattle occurs primarily in Mexico's tropical areas, is typically spread through the bites of vampire bats, and causes significant losses, particularly of young animals. In recent years, educational programs have helped producers identify the habitat of vampire bats (as opposed to the various other insectivorous bats in the region) and implement more effective control efforts, sharply decreasing rabies incidence in cattle in some regions.

Ticks (various species) are also widespread in cattle across Mexico, but tick loads (tick density on animal) are particularly heavy in tropical areas. Certain species of ticks also cause 'tick fever' (*babesiosis* or 'cattle fever'). Aside from direct disease impacts, heavy tick loads impact productivity. During wet periods, when tick infestations are highest, cattle must be treated every 10-14 days with an acaricide. Acaricide-resistant ticks are increasingly common, leading to greater concern about the effectiveness of current tick control measures (Peel, Johnson, and Mathews, 2010).

Mexican steers and heifers exported to the United States must be: [7]

- Properly identified;
- Retained in the export region for at least 60 days prior to shipment as long as that region remains disease free (i.e., not been quarantined for diseases in Mexico and inspected by an authorized veterinarian);
- Tested for tuberculosis and dipped for ticks prior to export; and
- Spayed (if a heifer), thus removing the ability to reproduce.

Cattle and Meat Marketing

In Mexico, cattle are typically purchased for feedlots directly from producers by buyer representatives. Private treaty sales are rarely reported, and there is no system for collecting or corroborating such private sales, especially for feeder cattle. Chihuahua and Sonora are the only States with sizable auction facilities, but they are not used to a great extent by feedlots in other regions. Cattle are usually purchased on a live-weight basis, with the pay weight determined on the truck near the source. Slaughter cattle and carcasses are reported more often, and a Federal price reporting system exists for larger cities.

Carcasses are priced on a pesos/kilogram basis and, increasingly, on a chilled basis. Historically, carcasses were marketed on a hot carcass basis, which in most instances included the hot carcass plus edible viscera (organs). While hot markets still exist, their numbers are dwindling. Differences in carcass price reporting are generally due to whether or not the byproduct drop value is included—the buyer gets only the carcass or gets the carcass plus the offal and hide. Viscera may also be marketed separately. Hides are often marketed separately by the cattle owner. In a small but increasing number of cases, meat is marketed in primary cuts (quarters, reflecting Spanish style cutting).[8] Primary cuts are currently sold in "full sets," essentially in carcass units. Nationwide, relatively little boxed beef production occurs at this time, with the exception of the Mexicali region, where roughly 90 percent of the beef is boxed, and areas of Hermosillo, Sonora, and Chihuahua.

Regional Summary

Differences in feed resources and cattle types are largely due to differences in geo-climates (fig. 6). Characteristics of major Mexican cattle feeding regions vary considerably as a result of geo-climatic differences and regional market preferences.

Mexicali, Baja California—The Mexicali area is a geographically isolated area that primarily serves Tijuana and the Baja California region. The Mexicali Valley includes a large area of irrigated land (by the Colorado River) used primarily for vegetable, wheat, and forage production. Wheat is the major feed grain used in feedlot production. Since few cattle are available locally, this area utilizes cattle from the farthest distance, on average, of any feeding area. Cattle fed here include No. 1 heifers and some steers from Sonora and Chihuahua, as well as bulls from the Pacific Coast. A limited number of cattle are sourced from all parts of Mexico and Central America.

The Mexicali region is geographically well defined and limited in size, but has a strong economic basis for the industry. Because of the geographical location and isolation, the industry is well situated relative to competition from the United States and other Mexican production areas. The high income level of the region provides a strong market basis for fed beef. The industry is one of the most technically advanced and represents the highest average level of intensity in stocker and feedlot production. The State of Baja California has the highest ratio of federally inspected slaughter plants, or "Tipo Inspección Federal" (TIF), to municipal slaughter of all Mexican States.[9] The area is well situated for possible exports to U.S. West Coast markets.

Source: Food and Agriculture Organization (FAO), 2005.

Figure 6. Geo-climatic regions of Mexico.

Monterrey, Nuevo Leon—The Monterrey area (including eastern Coahuila) is one of the oldest and most established cattle feeding regions. The close proximity of Monterrey to the United States has resulted in substantial merging of Mexican meat preferences with U.S. production styles, resulting in feeding that, in some cases, approaches the U.S. style, but still has less fat and marbling. In other cases, feeders in this region are focusing on the central Mexican market and are providing meat that is very lean. In all cases, however, there is a strong local preference for white fat and younger animals, which favors feeding heifers, while bulls are fed for markets farther south. Animal supplies consist principally of No. 1 heifers remaining from exports to the United States (mostly from Nuevo Leon and Coahuila) and No. 1 1/2 and 2 animals from the northern Gulf Coast region. Increasingly, animals from farther south are brought in when necessary to supplement deficits in local and regional supplies.

The State of Nuevo Leon has a mandatory meat grading system that is analogous to the U.S. system, but with different grade definitions. This is the only such system in use in Mexico. The four grades are Select (*Selecto*), Good (*Bueno*), Standard (*Estandard*), and Commercial (*Comercial*). The grading criteria are based on age (less than 42 months of age, based on ossification), lean color, fat coverage, and marbling. Most feeders achieve 80-85 percent Good, with less than 5 percent achieving Select. Commercial is used for older cows and bulls. Good, in this system, is roughly equivalent to Select in the U.S. system, while Select is close to low Choice in the U.S. system.

The Monterrey area is close to grain production in Tamaulipas (the largest grain sorghum producing State in Mexico), as well as sources of imported grain from the United States. The region has a relative shortage of forage. Few of the feeding operations include backgrounding or stocker operations, and there is more of a tendency to place cattle lighter and feed them for longer periods of time. Statistics for the State of Nuevo Leon indicate a considerable area of improved pasture, but apparently a substantial proportion of this is used for backgrounding calves for export and for cow-calf production.

The feeding industry in northeast Mexico has both advantages and disadvantages. Monterrey represents the third largest urban market in Mexico. It has a high general economic level and distinguishably different meat preferences from the rest of the country. Regional supplies of cattle and grain also support production. The area, in many ways, represents the economic boundary between U.S. and Mexican markets and is subject to strong and volatile economic forces. The region must compete simultaneously with the U.S. market for No. 1 cattle and with the threat of large quantities of imported meat. The Monterrey market is probably the most "Americanized" and thus is more vulnerable to encroachment from U.S. beef supplies. The possibility of increased exports of spayed heifers to the United States will increase competitive pressure for cattle sources in the area.

Tampico, Tamaulipas, and Huasteca—The southern part of Tamaulipas and the Huasteca region represent a developing area in the Mexican cattle feeding industry. Most of the feeding operations are new or have substantially expanded in the last 5 years. This development follows dramatic changes in production and marketing patterns in the area in the last 10 years. Until a decade ago, most cattle produced in the area (principally No. 2 type animals) were finished on grass. Cattle feeders from central Mexico then began purchasing locally produced calves for shipment to finishing areas. More pasture was available for cow-calf production as calves were marketed at weaning. Bulls that were traditionally castrated to improve fattening on grass were now left intact according to feeders' preferences for feeding bulls. (Steers in feedlots get too fat for Mexican preferences in most markets.) Producers soon began feeding cattle locally rather than shipping all weaned calves to other areas for feeding.

The feeding industry in the region is relatively small but growing rapidly, with a feeding capacity at 30,000-40,000 head. The area is close to grain, sorghum, and soybean production in Tamaulipas (about the same distance as Monterrey) but, like Monterrey, suffers from a relative shortage of forage. In the Tamaulipas and Huasteca regions, however, this shortage is more of a seasonal problem due to the dry tropical climate. A more common practice in the region is backgrounding calves with supplementation, but the region lacks quality pastures rather than insufficient quantities of forage. Cattle for feedlots come from a specific area, including the southern third of Tamaulipas, the northern third of Veracruz, and the Huasteca regions of San Luis Potosi, Queretero, and Hidalgo. Some Gulf Coast cattle from farther south are being placed in central Mexico as well.

This region is situated to expand considerably. The area has a clear focus on the markets of central Mexico and is well positioned to pull in more Gulf Coast cattle to be finished in feedlots rather than finished traditionally on pasture. The area is in some competition with the Monterrey area for feed grain, but this does not seem to be a serious problem for either area (both are close to U.S. grain sources, at least seasonally). The relative lack of forage in Huasteca is an issue of concern, but effective use of lower-quality forage with supplementation provides the needed flexibility to control production costs.

Jalisco, Guadalajara—Jalisco and Michoacan are traditionally among the strongest cow-calf production regions in the temperate climate areas of Mexico. For this reason, the area has been a major supplier of beef, not only for the Guadalajara market but for Mexico City as well. Historically, much of the beef was finished on pastures, but urban consumer markets are rapidly developing preferences for fed beef (although with very little finish).

Jalisco is part of the agriculturally rich Bajio region and is a major crop producing State. It ranks high in grain production—second in planted corn area (and the highest level of production), fourth in grain sorghum, and seventh in wheat. The State is also a major producer of forage crops and ranks first in planted area of corn and oat silage, sixth in sorghum silage, fifth in alfalfa hay, and second in area of improved pastures. The State is the largest producer of milk, pork, poultry meat, and eggs, and has the third largest inventory of beef cattle (Jalisco is the second largest producer of beef). Thus, the State has considerable competition for feeds among livestock industries but also substantial production of forages, lower quality feeds, and crop and food residuals that can be utilized by the beef industry.

Feedlot production has occurred in the region to a limited extent for many years but appears to be growing with changing preferences in the beef market. Cattle are generally fed for short periods (70-100 days) and marketed with very little external fat and little or no marbling. Many cattle are sourced in the Jalisco/Michoacan area, but some are brought in from farther south. Most heifers and bulls are placed after traditional backgrounding on pasture or intensive backgrounding on irrigated pastures in the Bajio region. Because of the short feeding period, cattle in this region are often placed in feedlots at somewhat heavier weights.

The Guadalajara region is the southernmost major concentration of feedlot production in the country, with large markets in both Guadalajara and Mexico City. The area has a clear focus on the differentiated beef preferences of central Mexico and on the types of cattle available in central and southern Mexico. The area has a generally favorable temperate climate and an abundant but diverse set of feed resources consisting primarily of forages of all qualities and some grain, but with a heavy emphasis on the use of residual and byproduct feeds and lower quality feeds not suitable for other livestock production. In recent years, cattle feeding has expanded in the Bajio, with cattle feeding in areas around Jalisco, including Aguascalientes, Queretero, and Michoacan. One of the newest and largest feedlots in Mexico was established in Michoacan by a major Mexican beef company in the last few years.

Comarca Lagunera (Torreon/Gomez Palacio)—The Lagunera region of southwest Coahuila and eastern Durango is a rich agricultural region built on irrigated crop production. Another traditional cattle feeding area, the region has undergone significant changes in recent years. Expansion in milk production, as well as poultry and pork production, in the region has decreased the relative role of fed beef production somewhat. This region includes diverse feeding operations, reflecting differences in production systems, cattle sources, and target markets.

Traditionally, the Torreon region evolved as a northern feeding area, much like Monterrey, feeding British/Continental and Zebu[10] cross cattle to meet northern preferences for meat with more finish and marbling. This is still the focus of some feeders in the areas that feed heifers and steers and produce meat principally for the Monterrey and Chihuahua markets. Torreon is well positioned in the central north to source cattle from the northeast in Coahuila and Nuevo Leon, as well as from the north and west in Durango and Chihuahua (fig. 7). However, competition with the U.S. market and other feeding areas in Mexico (along with drought-induced reductions in northern cattle inventories) has sharply reduced available supplies of northern cattle.In recent years, many feeders in the region have sourced cattle from the Gulf Coast and southern Mexico. Depending on the quality of the cattle, males may be fed as steers for the northern meat markets or, more likely, as bulls for the central markets. Cattle for central markets are fed shorter periods and with somewhat less intensity than cattle for northern markets.

As a result of increased dairy production in the region, some feeders are now specializing in feeding Holsteins and, most recently, a number of new back-grounding operations are developing Holstein calves in confinement prior to finishing. These operations are using residual and waste feeds from dairy operations to develop calves in confinement in the absence of pasture-based backgrounding programs.

Sonora, Hermosillo, Chihuahua—The State of Sonora has a large area of irrigated crop production and is the largest wheat, cotton, and safflower producing State in Mexico. Sonora also produces forage crops, including alfalfa hay and ryegrass pasture. Sonora has large pork and egg industries and significant cow-calf production.

The feedlot industry in Sonora is generally similar to other northern feeding areas, but the geographic position of the State is somewhat limiting. With one of the oldest feeding areas, the industry in Sonora appears to have lost some market position as feeding develops in Mexicali, Culiacan, and Torreon. With Mexicali positioned to serve the extreme northwest markets (Tijuana), and Culiacan and Torreon better positioned to serve other northern and central markets, Hermosillo has a distinct advantage only for relatively small regional markets and may play a mostly supplementary role for other markets. These limitations do not represent a threat to the industry as it exists, but it may represent something of an upper limit to development of the industry in the region. The same limitations may affect the region's ability to source cattle, with Mexicali and Torreon better positioned to attract northern cattle and Culiacan better positioned to attract cattle to the south. Sonora does, however, have a large base of irrigated ryegrass for backgrounding cattle in the dry winter period. The State of Chihuahua consists largely of dry, temperate, and semi-arid climate zones amenable to extensive grazing operations.

Source: Derrell Peel, Oklahoma State University.

Figure 7. Cattle feedlot in the Torreon region, Coahuila.

Cow-calf operations dominate the cattle sector, and Chihuahua accounts for over 40 percent of feeder cattle exports to the United States. Cattle from this and other nearby Mexican States are among the better quality cattle in Mexico and, with high percentages of English breeds (e.g., Hereford and Angus), finish well in feedlots.

Sinaloa, Culiacan—In many respects, the State of Sinaloa shares many crop production traits with its northern neighbor, Sonora. Like Sonora, Sinaloa has a large area of irrigated crop production. Because it has a predominantly dry tropical climate, however, the State has a larger component of temporal crop production than Sonora. Sinaloa has the third largest area of bean production but has the largest total bean production, due to larger irrigated yields. Sinaloa is the fifth largest grain sorghum producer and the fifth largest wheat producer in Mexico. Sinaloa is also the fourth ranking cotton producer. The State produces forage crops, including improved pasture and sorghum silage. Sinaloa has a large beef cow-calf industry and is a significant producer of pork and poultry meat.

Fed cattle production in Culiacan is heavily oriented toward central Mexican markets, utilizing predominantly Zebu-based cattle from the Pacific Coast (Sinaloa and Nayarit), as well as cattle from the Bajio region. The industry enjoys a relatively abundant and balanced set of feeds, including grains, protein feeds, and forages, and only moderate feed demand competition from pork and poultry.

Humid Tropics (Veracruz, Tabasco, Campeche, and Yucatan)—This large area presently accounts for little feedlot production, with perhaps capacity for 30,000-55,000 head. There are only a few large feedlots (5,000-25,000 head) in the region, including one in Merida and one in Veracruz, a handful of smaller feedlots (500-5,000 head), and numerous very small feedlots. However, larger feedlots have expanded significantly and limited cattle feeding now occurs in the tropical and southern States, including Oaxaca and Chiapas. Simultaneously, use of semi-intensive finishing programs thatcombine grazing and supplementation is increasing in the region. These changes are occurring in response to changing consumer meat preferences (for more fed beef) and increased demand from northern feeding areas for cattle for feedlots. Changes in production and marketing practices in the humid tropics may have significant implications for the total beef cattle industry in Mexico. Traditional grass-finished production in this region usually requires 2 years after weaning to finish cattle. Since 2005, several new production and marketing alternatives have arisen:

- Selling more calves at weaning for shipment to northern Mexico;
- Selling calves as feeders ready for placement in feedlots (local or northern);
- Semi-intensive finishing on pasture with supplementation; or
- Finishing in feedlots.

These alternatives reduce the post-weaning grazing period by 1-2 years and potentially release forage for increased cow-calf production. Because of the high stocking rates possible in this region, such changes could significantly alter the total productive potential of the beef cattle industry in Mexico.

The Mexican cattle and beef industry has been very dynamic for the last two decades. In the 1990s, prolonged drought coupled with the 1994 peso devaluation and subsequent limitations on capital access led to herd liquidation. Cyclically high U.S. cattle prices through much of the 2000s resulted in feeder cattle exports from Mexico at rates beyond long-term sustainability.

MEXICO'S CATTLE AND BEEF INDUSTRY POISED FOR GROWTH BUT FACES CHALLENGES

At the same time, growing beef demand and changing beef preferences by consumers led to increased beef imports and more cattle feeding in Mexico. The current strength of U.S. market prices and general strength of domestic beef demand have made it difficult to retain females for breeding. In recent years, restrictions on importing U.S. breeding stock due to animal health- and identification-related issues have also impeded herd rebuilding. Live breeding cattle exported to Mexico must be born after January 1, 1999 and are subject to animal identification requirements, all of which increase the total costs of importing U.S. breeding stock.[11] Continued growth in Mexican cattle exports to the United States is also vulnerable to issues of border security. Continued population and economic growth suggest considerable potential for increased fed beef demand, including demand for U.S. beef imports. The Mexican beef industry will face tremendous challenges to remain competitive in the face of:

- Product and feed demand from other livestock industries;
- Crop production for land and water to produce feed, especially forage; and

- Inputs (feed and cattle) and meat on the international market.

Although beef demand, specifically demand for fed beef, has increased significantly in recent years, demand for pork and poultry in Mexico has also increased. Poultry consumption has doubled, while beef and pork consumption has grown more modestly. An increase in total livestock production also means increasing competition with other types of food and fiber production for limited agricultural resources. These factors will become more critical as the beef industry adopts intensive production systems to produce fed beef and competes more directly with other livestock industries.

The change in demand for fed beef also has tremendous implications for the type of cattle fed in feedlots, feeding strategies, and the quantity of fed cattle production. Moreover, a change from grass-fed beef to fed beef in Mexico has major implications for overall levels of beef production. Domestic production is limited by forage availability for the cow herd, which is currently limited by forage used for finishing beef on pastures. An increase in more intensive stocker and finishing systems will release forage resources for increased cow-calf production. Changing beef consumption patterns in Mexico also will impact marketing. Changes in retailing, especially rapidly expanding supermarkets and growth in away-from-home consumption, are part of a large set of lifestyle changes causing significant shifts in market infrastructure and marketing practices within the industry.

Acknowledgments

We owe much to the reviewers who took their time to provide us with helpful comments and insights. Thanks to Economic Research Service reviewers Don Blayney and Keithly Jones, Foreign Agricultural Service reviewers Claire Mezoughem and Zaida San Juan, and New Mexico State University reviewer Rhonda Skaggs. The report was cleared by Shayle Shagam at USDA's World Agricultural Outlook Board. Special thanks also to our editor, Angela Anderson, and layout specialist, Vic Phillips, Jr.

About the Authors

Derrell S. Peel is the Breedlove Professor of Agribusiness and an extension livestock marketing specialist at Oklahoma State University. Kenneth H. Mathews, Jr. and Rachel J. Johnson are agricultural economists with USDA's Economic Research Service.

References

Batres-Marquez, S.P., R.L. Clemens, and H.H. Jensen. "Mexico's Changing Pork Industry: The Forces of Domestic and International Market Demand," *Choices* 22(1), 2007.

Hale, D.S., K. Goodson, and J.W. Savell. "Beef Quality and Yield Grades," Department of Animal Science, Texas AgriLife Extension Service, 2010.

Mejia, M., and D.S. Peel. "White and Yellow Corn Production in Mexico: Food versus Feed," *Analysis and Comments*, Letter No. 25, Livestock Marketing Information Center, Denver, CO, June 2009.

Mendez, R.D., C.O. Meza, J.M. Berruecos, P. Garces, E.J. Delgado, and M.S. Rubio. "A survey of beef carcass quality and quantity attributes in Mexico," *Journal of Animal Science* 87: pp. 3782-90, 2009.

Peel, Derrell S., Rachel J. Johnson, and Kenneth H. Mathews, Jr. "Cow-Calf Beef Production in Mexico," LDP-M-196-01, U.S. Department of Agriculture, Economic Research Service, November 2010, http://www. ers. usda.gov/Publications/ LDP/2010/10Oct/LDPM19601/.

U.S. Department of Agriculture, Animal and Plant Health Inspection Service. *Animal Import*, http://www.aphis.usda.gov/import_export/animals/live_ animals.shtml.

U.S. Department of Agriculture, Foreign Agricultural Service. *Livestock and Poultry: World Markets and Trade*, October 2009.

End Notes

[1] Continental and British breeds are two categories of Bos taurus breeds of cattle. While both categories originated in Europe, the British breeds, specifically originating from the British Isles, are smaller in size, mature early, and have increased fleshing and marbling ability. The larger Continental breeds are leaner, more muscular, and, as with the British breeds, vary in adaptability to hot climates.

[2] For the purposes of this report, "fed cattle" refers to cattle finished on grain rations and is equivalent with the term "grain-fed." Fed cattle in Mexico are generally fed a smaller amount of grain in the ration compared with cattle fed in the United States.

[3] "Semi-intensive" production systems are those in which cattle are not generally confined during the growing or finishing periods. "Intensive" production is used in an environment of close confinement, like backgrounding lots or feedlots.

[5] Muscle thickness is the ratio of muscle to bone at a given degree of fatness and is measured in grades. Three muscle thickness grades are currently used, No. 1, No. 2, and No. 3, from thickest to thinnest, respectively.

[6] Cattle sourced from Central America must be in Mexico at least 60 days before they are eligible for export to the United States, according to bovine import protocol set by the USDA, Animal Plant Health Inspection Service (APHIS), http://www.aphis.usda.gov/ import_export/ animals/live_animals. shtml.

[7] USDA, APHIS import protocol for cattle from Mexico can be found at http://www.aphis. usda.gov/import_export/downloads/pro_imp_bo_feed-ers_mx.pdf.

[8] Styles of cutting beef carcasses vary from country to country or region to region. Spanish-style cuts are oriented more for grilling and slow cooking.

[9] TIF slaughter plants have the highest sanitary standards and are the most advanced in technological processing. The plants are certified and inspected by Secretaria de Agricultura Ganaderia, Desarrollo Rural, Pesca y Alimentacion (SAGARPA). TIF plant services include slaughtering, carcass handling, packaging, refrigerated storage, and fabrication of processed products. Municipal plants offer limited services, primarily slaughtering and carcass cutting. They are not required to follow sanitary control measures and often lack the equipment and resources necessary to properly dispose of byproducts (Batres-Marquez, Clemens, and Jensen, 2007).

[10] Zebu (Bos indicus) breeds used in Mexico include Brahman, Nelore, Gir, Indubrasil, and Guzerat. These breeds share a number of phenotypic characteristics, including the distinctive hump and pendulous hide. However, each of these breeds can be distinguished by color patterns, horn shape, and ear configuration.

[11] USDA, APHIS guidelines for exporting U.S. breeding cattle to Mexico can be found at http:// www.aphis. usda.gov/import_export/downloads/ mx_bo_breeding.pdf.

In: There's The Beef
Editors: Jonathan L. Ruiz and S. R. Fox

ISBN 978-1-61942-943-7

Chapter 6

COW-CALF BEEF PRODUCTION IN MEXICO*

Derrell S. Peel, Rachel J. Johnson and Kenneth H. Mathews, Jr.

ABSTRACT

Mexico often exports over a million head of cattle annually to the United States. These animals, primarily intended for U.S. stocker and/or feeder operations, all pass through a cow-calf production system of one form or another in Mexico. This report outlines cow-calf production types, beef production systems, and geo-climatic influences on beef production in Mexico. The authors of this report also evaluate factors affecting Mexican beef production, such as patterns of landownership, major diseases and parasites in cattle production, changes in export regions, and changes in domestic consumption as they pertain to cow-calf production. Amid changes in consumer demand and the continuous challenges pertaining to Mexico's cattle export market, production practices at the cow-calf level may pinpoint opportunities and constraints within the domestic beef markets of both Mexico and the United States.

Keywords: Cow-calf, Mexico, beef production, cattle

INTRODUCTION

Cattle production provides vital economic activity for the large expanse of nonarable land in Mexico, and the United States is the primary export market. Cattle raised for export in Mexico represent, on average, more than half of all U.S. cattle imports. In 12 of the years between 1989 and 2009, Mexico exported over a million head of cattle, mostly steers and heifers for feeding, to the United States (fig. 1).

---

* This is an edited, reformatted and augmented version of The United States Department of Agriculture publication, LDP-M-196-01, a report from the Economic Research Service, from www.ers.usda.gov, dated November 2010.

Exports of Mexican cattle fluctuate seasonally, depending on weather conditions and the availability of forage for grazing. Cattle exports tend to peak in the early spring (February-March) and late fall (October-November), as forage conditions deteriorate (fig. 2). Lighter weight feeder and stocker cattle[1] are exported since the United States has the comparative advantage in feeding cattle and growing cattle on available forages.

Recent shifts in cattle-exporting regions reflect longer term changes in the industry. Improved quality, combined with improved health status in several regions and a generally strong demand for Mexican cattle in the United States in the past decade, has led to increased exports from more Mexican States. Mexico's ability to remain a strong cattle exporter to the United States, however, depends principally on the health and quality of the cattle and breed characteristics. Cattle intended for export to the United States typically originate from cow-calf operations in Mexico as stocker/feeder animals. Improvements in cattle quality and management practices at the cow-calf level will increase the number of cattle that meet changing consumer demand in Mexico and are suitable for export to the United States.

Geo-Climatic Production Regions

Mexico is divided into four climatic areas: arid, semi-arid, temperate, and tropical (fig. 3). Tropical areas can be further divided into humid and dry tropics. Cattle are produced in virtually all of Mexico's climatic environments (fig. 4), with production systems and costs dependent on climate. The productivity of each cattle production system—or costs to achieve a given level of productivity—varies from region to region.

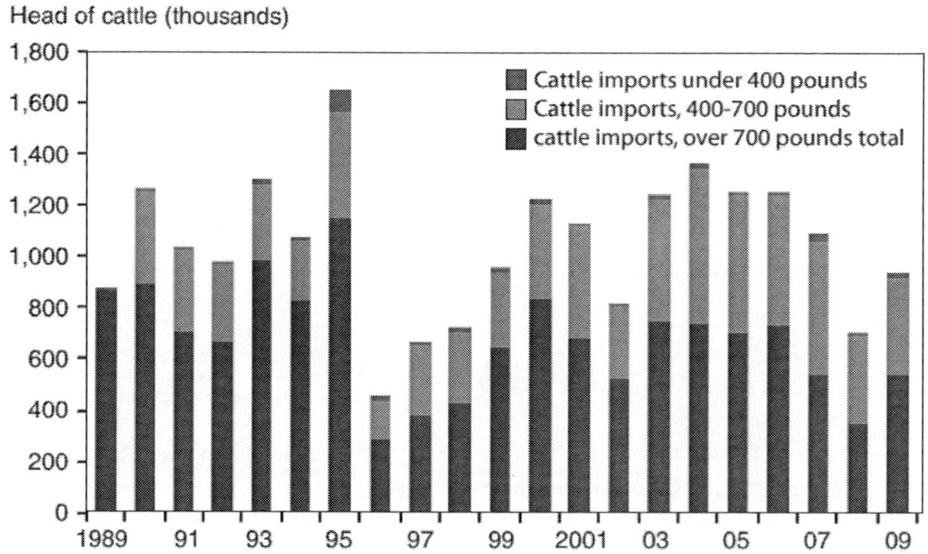

Source: USDA, Economic Research Service.

Figure 1. Cattle exported from Mexico to the United States, 1989-2009.

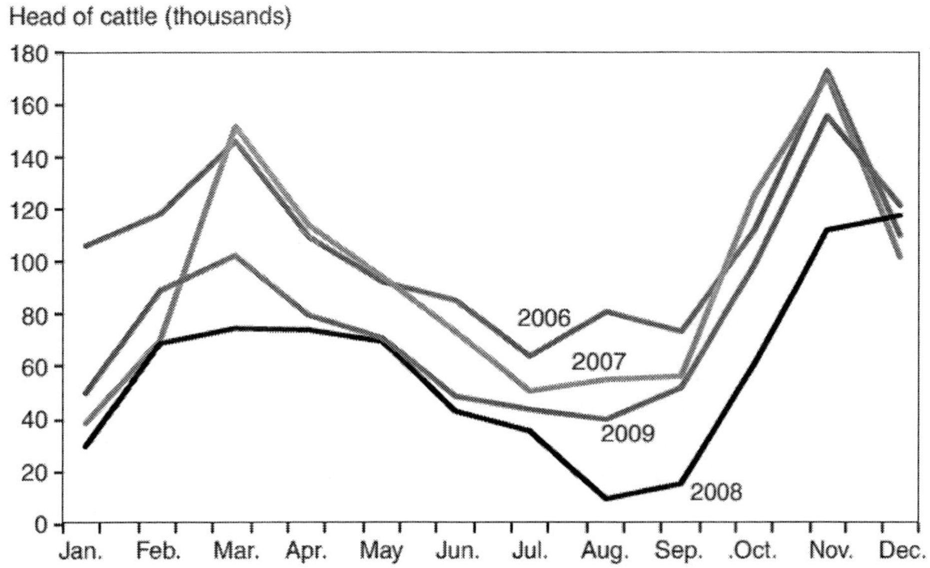

Source: USDA, Economic Research Service.

Figure 2. Cattle exports from Mexico to the United States follows a seasonal pattern.

Arid

Approximately 20 percent (about 41 million hectares) of Mexico is arid. Most of the arid land is located in the States of Baja California, Baja California Sur, Sonora, Chihuahua, and Coahuila, with smaller proportions in Durango, Nuevo León, San Luis Potosi, and Zacatecas. Arid lands range from desert areas completely devoid of vegetation to areas of desert scrub (chaparral) dominated by shrub species with some grass and forbs.[2] Some of these chaparral areas contain plant species of good palatability and nutritional quality, but forage production is low. The four primary cow-calf production systems in Mexico (described later) can be adapted to arid regions. In addition to water, one of the principal limitations to production in arid regions is the extensive area needed for grazing. Stocking rates are very low in arid areas, ranging from 25 to more than 50 hectares per animal unit,[3] and supplemental feed is essential for more intensive production. The arid climate also favors a moderate cow size. The large Continental[4] breeds, with higher daily energy requirements than smaller breeds, may be challenged to find sufficient forage. The breeds that do best in the arid regions of Mexico are the smaller British breeds, which may have some Zebu[5] cross.

Semi-arid

Semi-arid climate is found in about 28 percent of Mexico, adjacent to the arid areas in the North and stretching farther down the central mesa. The highest proportions of semi-arid land (to total State area) are found in Aguascalientes, Chihuahua, Guanajuato, Hidalgo, Oaxaca, San Luis Potosi, Tamaulipas, and Zacatecas.

Smaller proportions are found in Sonora, Durango, México, Puebla, Coahuila, and Jalisco. Semi-arid areas typically include short-grass prairies with some mixture of forbs and shrubs (fig. 5).

Like the arid regions, forage production on semi-arid land is seasonal and characterized by a short rainy season in the summer and fall, followed by a 6- to 7-month dry season. Forage quality remains relatively good in the dry season, but careful pasture management is required to ensure adequate quantities of forage. Stocking rates range from 12 to 25 hectares per animal unit.

Source: Food and Agriculture Organization (FAO), 2005.

Figure 3. Geo-climatic regions of Mexico.

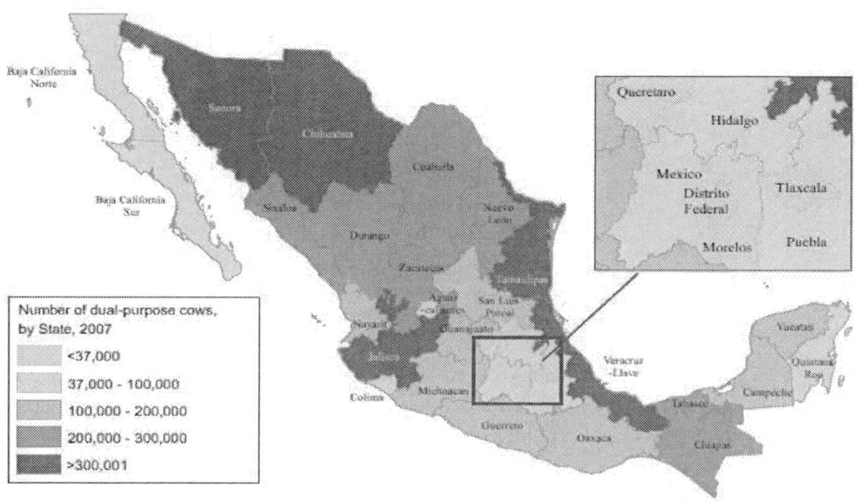

Source: INEGI Censo Agropecuario, 2007.

Figure 4. Beef cattle are not uniformly distributed among Mexican States.

Source: Derrell Peel, Oklahoma State University.

Figure 5. Production of VacaNorte cattle on semi-arid rangeland in Chihuahua.

A protein supplement is usually required in the dry season as well as an energy supplement when forage quantity is limited. Cow size is not as limited as for arid land, but the reproductive performance of large Continental breeds will decrease quickly with nutritional shortages. Many large operations producing cattle for export to the United States are located in the arid and semi-arid States of Mexico.

Temperate

Mexico's mountainous interior accounts for 23 percent of the area characterized by a temperate climate. Temperate climates are widespread, with nearly every State having some temperate area, and extend from the Sierras in Chihuahua to the highlands of Chiapas. Most of the temperate climate area is in the interior part of central Mexico. High proportions of temperate climate areas are found in Tlaxcala, México, Michoacán, Guerrero, Jalisco, and Hidalgo. The temperate climate region is divided between forested mountains and cultivated valleys.

Beef production plays a relatively small role in this region compared with crop production and other livestock production activities. Also home to over 60 percent of the Nation's population, large quantities of dairy, pork, poultry, and egg production occur in the temperate region. Compared with the other climatic regions, many beef operations in the temperate region are small and are operated in conjunction with farming and other activities. Large tracts of extensive range are not common. Many cattle are produced with relatively little permanent pasture and require large proportions of crop residues and byproduct feeds. Stocking rates in the temperate region range from 4 to 12 hectares per animal unit. Production costs are moderate in the region, but productivity is often low because of generally fewer management practices. Climate, disease, and pest problems are less severe in the temperate areas, but the annual dry period leads to nutritional stress that Zebu cattle are better suited to sustain than British or Continental breeds, especially in the more southern States of Mexico.

Dry Tropics

Dry tropics make up about 16 percent of the total land area of Mexico. Dry tropical regions are characterized by annual frost-free temperatures and high rainfall, but they also experience a seasonal arid pattern. Dry tropics are distributed widely across Mexico, with large concentrations in the Yucatan Peninsula, including the States of Yucatán and Campeche; the northern Gulf Coast, including northern Veracruz and Tamaulipas; and much of the Pacific Coast, including parts of Sinaloa, Nayarit, Colima, Guerrero, Oaxaca, and Chiapas.

The dry tropics provide a challenging beef production environment due to the heat, disease, and pest factors common to tropics but with the added burden of a dry season, which results in reduced forage quality and quantity. The stocking rate in dry tropical regions is generally 3-6 hectares per animal unit. All calf production practices are possible, but production of higher quality cattle for export means increased costs for feed and pest/disease control. In the tropical areas of Mexico, the climate and production environment strongly favor the use of Zebu cattle.

Humid Tropics

The humid tropics are characterized by high annual rainfall and green vegetation and account for the remaining 13 percent of Mexico's land area. Biomass production is high, and thus, stocking rates are high, ranging from 1 to 3 hectares per animal unit. Forage quality is relatively low, resulting in lower animal productivity unless supplemental feed is provided. Heat and humidity, combined with severe disease and pest problems, make Zebu cattle preferable in this region.

Major Mexican Cattle Diseases and Parasites

Cattle producers in Mexico face a wide variety of production challenges, including diseases and pests. Aside from significant economic and trade impacts, several of the major Mexican cattle diseases and parasites also represent serious human health threats.

Bovine Tuberculosis (TB) is the primary disease that limits Mexico's access to export markets for cattle. TB is endemic in the country, though reasonably well controlled in most areas. Mexico has supported a Federal TB eradication program since 1971 and, in conjunction with individual States, has made substantial progress, particularly in the northern States, where cattle production is often focused on calves for export. TB remains widespread in concentrated areas of dairy production in the northern States, threatening the status of surrounding areas of beef cattle production. In other parts of the country, isolated regions have achieved a TB status that permits cattle exports, but that status is difficult to maintain because of surrounding endemic areas. Although TB affects the productivity of cows for meat or for milk production, the effects are so subtle that many producers fail to recognize the impacts, making voluntary cooperation in eradication programs difficult. TB is a zoonotic disease that threatens human health as well.

Brucellosis is currently endemic and widespread in Mexican cattle. The prevalence of brucellosis prevents exports of breeding animals, but its biggest impact is on productivity. The disease reduces reproductive performance but, like TB, usually does not affect the overall health of infected animals. The principal impact of the disease is spontaneous abortion in cattle; since the disease often goes undetected or undiagnosed, the disease's impacts are often underestimated. In addition to the Federal TB eradication program, Mexico also has a Federal brucellosis eradication program, but it requires greater resources and progress in implementation has been slow. Brucellosis is also a zoonotic disease, threatening humans, primarily through consumption of unpasteurized milk and dairy products.

Rabies, also a zoonotic disease, occurs throughout Mexico and affects a wide range of wildlife and domestic animals (especially dogs), as well as people. Rabies in cattle occurs primarily in Mexico's tropical areas and is typically spread through the bites of vampire bats, often resulting in significant losses, particularly of young animals. In recent years, educational programs have helped producers identify the specific habitat of vampire bats (as opposed to the various other insectivorous bats in the region) and implement more effective control of the bats, sharply decreasing the rabies incidence in cattle in some regions.

Mexico is host to several species of *ticks*, including those that cause tick fever (*babesiosis* or cattle fever). Ticks are widespread in cattle across Mexico, but tick loads (tick density on animal) are particularly heavy in tropical areas. Aside from direct disease impacts, heavy tick loads impact productivity. During wet periods, when tick infestations are highest, cattle must be treated every 10-14 days with an acaricide. Acaricide-resistant ticks are increasingly common, leading to greater concern about the effectiveness of current tick control measures.

THE MEXICAN LANDSCAPE AND COW-CALF PRODUCTION

Mexico's cow-calf industry encompasses a wide array of production activities that vary according to a diverse set of economic motivations, political and legal environments, and cultural adaptations. Cow-calf enterprises produce meat, milk, breeding stock, animals for work, and animals for recreation, including fighting bulls and rodeo cattle. These diverse enterprises are supported by beef production in climatic conditions ranging from tropical to desert. Given these many factors, Mexican cow-calf production is classified by product, by region, by management system, and by animal type.

Mexico offers extensive forage resources, and over half of its 196 million hectares are used for livestock production. Cattle consumed the most forage in 2007, with a total inventory of over 23 million head, but 9 million head of swine, 7 million head of sheep, and 4 million head of goats also compete for forage resources (table 1). Forage resources include forbs and shrub species in arid regions, semi-arid and temperate native range that extends into forested areas, and tropical grazing areas, some that are available only seasonally due to the presence of standing water.

Table 1. Mexican livestock inventory, by species and type, 2007

Species/type	Number (head)
Total cattle	23,316,942
Total swine	9,021,192
Sows	1,070,716
Total sheep	7,305,578
Ewes	1,886,856
Sheep for wool	2,724,157
Total goats	4,124,201
Does	1,140,297
Milking goats	323,034
Total equine	2,143,934
Horses	1,328,524
Mules	234,009
Donkeys	581,401
Total poultry	356,824,337
Hen chickens	118,019,090
Broilers	159,946,197

Source: 2007 Censo Agropecuario.

Cattle producers also make extensive use of crop residues and byproduct feeds, including sugarcane bagasse, citrus pulp and waste, brewer's grains, and discarded food products.

Mexico's beef industry has faced minimal competition from the crop production sector given the country's limited availability of arable land. Traditionally, most beef was forage-finished and utilized forages from nonarable lands, and the feeds produced on cropland were mostly alfalfa and feed grains used for dairy, poultry, and swine production.

In the last decade, a rapidly growing consumer preference for fed beef (animals are fed in feeding operations similar to U.S. feedlots and produce a higher-quality beef carcass) has resulted in increasing demand for feed grain and high-quality forage resources by the beef industry. This relatively new demand for cultivated feeds—combined with growth in dairy, pork, and poultry production—has intensified the competition between food grains, feed grains, and forage production in Mexico's crop production sector (Mejia and Peel, 2009). Amid the potential for rising feed costs, increased competition in the livestock sector, and continued changes in consumer demand for beef, improvements in cow-calf management and production practices may determine the strength of Mexico's beef industry, as well as that of its cattle exports.

LAND OWNERSHIP AND LIVESTOCK PRODUCTION

Based on a complicated tenure system, Mexico's land can generally be divided into two sectors: private and social. Private lands are what remain of the haciendas that existed at the beginning of the 20th century. Most of the ranches that form the commercial beef cattle industry are known as "pequeña propiedad," or small land holdings. Nearly 80 million hectares, just under 40 percent of Mexico's arable land area, is considered private land. In

2007, over 105 million hectares remained in the social sector, with about 32 percent, or 33.6 million hectares, in parcels and most of the remaining portion (69.3 million hectares) in common land, or "ejidos" (INEGI). Although Article 27 of the Mexican Constitution was modified in 1992 to establish a process for adjudicating ejido land ownership rights so the land could be bought, sold, leased, or mortgaged, common lands still account for nearly 60 percent of the land used for livestock production in Mexico.[7]

Historically, ejido lands have been managed differently than private lands and, in many cases, the ejido cattle production units are much smaller than the private ranches. Traditional ejido agricultural production has often focused on crop farming, with livestock production treated as a complementary or residual activity, resulting in less management intensity in livestock production compared with the private sector. In many instances across Mexico, ejido lands have been overutilized for livestock and forest production, deteriorating quality and productivity (Cartron, Ceballos, and Felger, 2005). Ejido beef production is also often focused on subsistence production using crop residuals and byproducts as livestock feeds.

The inability to sell or lease ejido land has greatly reduced the cost or opportunity cost for the use of grazing resources. The observed differences in productivity and economic costs make questions regarding the use, management, and productivity of these lands important for the overall potential of the beef industry. The distinction between ejido and private land may decrease over time, at least regarding productivity and costs of production, but only in regard to the parcelized ejido lands. The questions surrounding the disposition of the common ejido lands will continue to affect Mexican cattle production until the issues can be resolved.

THE MEXICAN CATTLE INDUSTRY: BY THE NUMBERS

Table 1 provides an overview of the inventory of livestock by species according to Mexico's 2007 Census of Agriculture. The census reported 23.3 million head of cattle in Mexico on some 1.13 million production units (farms and ranches), resulting in an overall average herd size of 21 head. Roughly a third of production units had fewer than 5 head of cattle, and no additional information was collected on these units. The remaining two-thirds of the production units had a total inventory of 22.1 million and an average herd size of 29 head. The total cow herd in Mexico was 8.7 million, of which 37 percent were beef cows, 34 percent dairy cows, and 28 percent dual-purpose beef and dairy cows (table 2).

The census identified cattle for three quality categories: "corriente," crossbred, and purebred. Within the purebred category there was also a designation for registered purebred animals. Corriente, in this case, means typical or common and refers to animals of mixed breeding with unimproved genetics.[8] Corriente cattle vary by region in Mexico and may include Spanish genetics, but usually consist of various nondescript beef, dairy, and/or Zebu breed influences. Corriente cattle are typically designated as dual-purpose, but may include some beef cattle as well. Roughly 16 percent of Mexican cattle are classified as corriente (table 2). Most of the 5.2 percent of (backyard) cattle not included in the classification may also be corriente, making the overall total 20 percent or higher. Crossbred cattle account for 52 percent of all cattle in Mexico and include most of the dual-purpose cattle and some beef cattle. Crossbred breed types vary regionally: those in the northern States include British and Continental breeds, often with some Zebu influence, while in the central and tropical areas most are dual-purpose cattle that are crossbred dairy and Zebu breeds. Purebred animals represent 27 percent of the total cattle inventory, with nearly 5 percent of the total registered as purebred animals. This category includes most dairy animals and some beef animals. Purebred dairy animals are typically Holsteins, and the purebred beef animals include at least some representation of most beef breeds.

Table 3 lists the top 10 States for total cattle inventory, total cow herd, beef cows, and dual-purpose cows, highlighting some important regional differences in cattle production in Mexico. Dual-purpose cows are located primarily in the south and central regions, plus coastal areas of the northwest. The top States for dual-purpose cows include Veracruz, Jalisco, Chiapas, Sinaloa, Guerrero, and Sonora. Beef cows are typically located in the north, including the States of Chihuahua, Sonora, Tamaulipas, Nuevo Leon, and Coahuila. Jalisco is the only major beef cow State in the central region.

Table 2. Mexican cattle inventory, by class, age, breeding, and management system, 2007

Category	Total (head)
Total	23,316,942
<5 head/production unit[1]	1,207,313
Class	
Bulls	616,820
Cows	8,671,516
Dairy	2,966,117
Beef	3,238,922
Dual-purpose	2,466,477
Feeders/on-feed/replacements	12,571,993
Work (tracción)	249,300
Age	
<12 months	6,732,250
1-2 years	4,296,529
2-3 years	3,897,912
>3 years	7,182,938
Breeding	
Corriente	3,712,561
Cross-bred	12,210,003
Purebred	6,187,065
Registered	1,121,336
Management system	
Free range	12,885,647
Controlled grazing	4,437,930
Confined	3,083,930
Semi-confined	1,702,534

[1]Production units reporting fewer than 5 head of cattle–representing 5.2 percent of the total cattle inventory—were not questioned about class, age, breeding, or management system. The remainder of the table is based on a total cattle inventory of 22,109,629 head. Source: 2007 Censo Agropecuario.

Table 3. Top 10 Mexican cattle producing States

Total cattle		Total cows[1]		Beef cows		Dual-purpose cows	
State	Head	State	Head	State	Head	State	Head
Veracruz	2,454,171	Veracruz	1,040,769	Chihuahua	363,336	Veracruz	428,554
Jalisco	1,931,546	Jalisco	949,252	Sonora	289,573	Jalisco	252,820
Chihuahua	1,708,887	Chihuahua	625,987	Tamaulipas	243,742	Chiapas	175.188
Chiapas	1,406,419	Sonora	507,795	Jalisco	242,097	Sinaloa	162,435
Sonora	1,351,642	Chiapas	489,321	Nuevo Leon	223,850	Guerrero	149,208
Durango	1,232,525	Durango	459,491	Coahuila	221,001	Sonora	134,018
Tamaulipas	1,054,831	Sinaloa	400,235	Veracruz	204,944	Zacatecas	125,493
Michoacan	1,004,538	Coahuila	389,312	Durango	187,936	Durango	111,574
Sinaloa	964,712	Michoacan	345,208	Tabasco	140,069	Michoacan	109,589
Tabasco	957,761	Guerrero	341,934	Sinaloa	118,651	Oaxaca	97,820

Source: 2007 Censo Agropecuario; INEGI.
[1]Includes specialized dairy cows.

The Mexican cattle production system relies on open range, controlled grazing, confined production, and semi-confined production. Nationally, 55 percent of all cattle are produced using open range (or open grazing) systems. Open grazing tends to be more important in the northern States and is associated with beef cattle production.

Controlled grazing accounts for 19 percent of cattle production. Confinement is used typically in dairy production and beef feedlots, accounting for 13 percent of cattle production. Semi-confinement is used for some dairy and dual-purpose production and accounts for 7 percent of cattle production.

The reported age distribution of the 2007 cattle inventory includes 6.7 million head of cattle that are younger than 12 months of age. Using this number as an estimate of the calf crop and dividing it by 8.7 million cows suggests an average calf crop of 78 percent. This estimate may overstate the actual calf crop percentage if the calf inventory includes calves from heifers not included in the cow inventory.

More importantly, the calf inventory also likely includes calves from cows that were not included in the cow inventory, such as "backyard" cattle or residual cattle kept by subsistence farmers. Adjusting the calf crop estimate to include some proportion (ranging from 50 to 80 percent) of backyard cattle reduces the estimated calf-crop percentage to a range of 70 to 73 percent.[9]

Additionally, it is reasonable to assume a significantly higher calf-crop percentage for dairy cattle, and a further adjustment using an assumed dairy calf-crop percentage provides an estimate of the nondairy (beef plus dual-purpose) calf crop percentage. For example, using an assumed 90 percent calf crop for dairy cattle (with the backyard cattle adjustment) results in an estimated nondairy calf-crop percentage of 61 percent. The true nondairy calf-crop percentage is likely between 60 and 67 percent as a national average. The average undoubtedly varies by region, ranging from as low as 50 percent in some regions to perhaps 75 percent in others.

Cattle Production Technology and Management

The 2007 census also provided information about the production practices and technology used in Mexican cattle production. Information was reported by cattle numbers and/or production unit. For example, 90 percent of all cattle (on 92 percent of production units) received vaccinations. Other cattle production practices of note include:

- 2 percent of all cattle (on 82 percent of production units) were treated for parasite control, mostly dipping vats (79 percent);
- 28 percent of all cattle (on 25 percent of production units) were given a balanced ration;
- 4.6 percent of all cattle (on 4.3 percent of production units) were impreg-nated via artificial insemination; and
- 0.35 percent of all cattle (on 0.6 percent of all production units) were
- impregnated via embryo transfer.

For production units only, the following information is provided:

- 60 percent used salt/minerals;
- 16 percent used a controlled breeding season;
- 32 percent used pasture rotation;
- 3.9 percent used technical assistance; and
- 1.8 percent participated in the Federal genetic improvement program.[10]

Types of Beef Cow-Calf Production

Three main commercial beef cattle production activities exist in Mexico: cattle for export, domestic beef production, and dual-purpose beef and dairy production. These categories ignore specialized milk production (dairies, mostly confinement operations), purebred beef production, draft animals, and specialty activities, such as production of rodeo cattle and fighting bulls. These activities contribute meat to the beef industry secondary to their principal focus.

Cattle for Export

Cattle for export (mostly to the United States) are produced mainly in the northern third of Mexico on larger ranches and increasingly are being marketed directly to U.S. buyers from the ranch. Cattle for export are primarily of British/Continental breeding, generally with no more than three-eighths Zebu influence. Historically, steers were exported to the United States, while heifers remained in Mexico to be sold into the domestic market either for meat production or as breeding stock. Recently, the increased feasibility and lower costs of spaying heifers are such that, when U.S. cattle prices are cyclically high, a substantial number of heifers are also being exported. All female cattle destined for stockering and feeding must be spayed before entering the United States.[11]

Tamaulipas and Chihuahua has traditionally been the largest cattle-exporting State (table 4). Chihuahua and Coahuila together provide about 72 percent of total Mexican cattle exports to the United States. Export origins, however, have changed over time and can be disrupted by regional health status. For example, in previous years, the State of Durango accounted for roughly 15 percent of Mexican cattle exports, but more recently has averaged less than 2 percent due to health restrictions. When traditional export cattle sources are restricted, other regions make up some of the difference.

Recently, exports have increased from northeastern Mexico, including Tamaulipas and northern Veracruz. Cattle from northeastern Mexico have generally been of somewhat lower quality than those from traditional exporting regions. However, a general improvement in cattle quality in many regions has narrowed the differences in quality; as a result, more cattle are suitable for export now than in the past.

Domestic Beef Production

Much of the beef production in the temperate and tropical areas of Mexico is destined for the domestic market. The pronounced differences between the domestic beef industry and the export industry arose partly for historical reasons and partly due to lack of proximity to the U.S. market. Mexican climate and management systems tend to favor a higher use of Zebu genetics, resulting in cattle less desirable to U.S. markets. Historically, calves from both beef and dual-purpose cattle systems were grown and finished on pasture, producing a traditional grass-fed beef. Today, many of those calves move into a semi-intensive or confinement finishing system to produce fed beef for the domestic market.

Table 4. Major Mexican cattle-exporting States, 2006-2008

State	Percent of total exports
Chihuahua	43.40
Tamaulipas	17.03
Coahuila	11.26
Sonora	6.95
Nuevo Leon	4.95
Zacatecas	4.38
Sinaloa	2.08
Durango	1.85
Campeche	1.06
Other States	7.04
Total	100.00

Source: INEGI, 2006-08.

Dual-Purpose Production

As the name suggests, dual-purpose operations produce both beef and milk. One of the advantages of dual-purpose production is the ability to emphasize either beef or milk production, depending on prevailing market conditions. Most notably, in the temperate and tropical regions of Mexico, dual-purpose production is accomplished with crossbred Zebu and dairy-type cattle, traditionally Brown Swiss. For example, some producers maintain dual-purpose cows that are well adapted to the local environment, but use beef breeds to produce a beef-oriented terminal cross calf, or animal that will not be used for breeding purposes and genetic improvement. In contrast, Zebu cattle are crossed with Holstein or other dairy breeds for dual-purpose production when milk production is emphasized over beef production (fig. 6). In some regions, formerly dual-purpose cattle producers are specializing more in either beef or dairy production.

Backyard Cattle Production

"Traspatio" or backyard animals are not a significant component of the commercial beef cattle industry in Mexico, but they represent a sizable potential for improvement and increased production. The backyard sector complicates production for the domestic beef market because it is difficult to document how traspatio production contributes to total slaughter rates and rural beef consumption. It is useful, however, to consider the types of animals included in this group and their role. As previously noted, 1.2 million head of cattle (5.2 percent of the total cattle inventory in 2007) are in production units with fewer than 5 head. The census figures imply that these farms have an average of 3.2 head per production unit.

Source: Derrell Peel, Oklahoma State University.

Figure 6. Dual-purpose cow raised for domestic meat/milk production in Veracruz.

Most of these cattle are part of family herds for subsistence farmers, and the majority of the cattle are kept for milk production or possibly as draft animals. In many cases, beef production is not the primary focus, and excess or cull animals are used for home consumption only occasionally.Backyard animals are not primarily used for meat production, chiefly because beef animals require large amounts of feed resources and a rather lengthy time to mature before slaughter. Subsistence farmers are likely unable to use beef carcasses for family meat supplies as refrigeration is often unavailable. Such families often rely on smaller livestock species (sheep, goats, swine, and poultry) for meat. Thus, a substantial proportion of backyard cattle are likely females and the calves produced are sold as calves.

States with the highest proportion of backyard production animals that were excluded from the census include some States with relatively few animals, such as Tlaxcala and the Federal District (Mexico City). Several other States, however, including Mexico, Oaxaca, and Veracruz, have significant numbers of backyard cattle.

Cow-Calf Production Systems

Based on management and product considerations, four Mexican cow-calf production systems can be defined and profiled: VacaNorte, semi-intensive, traditional, and Criollo. Production occurs in a continuum of systems, but these classifications generally cover the range and principal differences in types of production. Table 5 compares the production characteristics of each system. The different classes of production are not explicitly tied to regions or climates, and all are found to varying extents in all locations. However, some production classes tend to be heavily associated with particular geographic regions.

Calves for export to the United States are a principal output of "VacaNorte" production system. Products include No. 1 calves (generally British or Continental crossbred animals with less than three-eighths Zebu influence) and No. 11/2 calves (British or Continental cross with up to one-half Zebu influence). As the name suggests, this system tends to be used mostly in the arid/semiarid northern part of Mexico, but it can also be used throughout the temperate part of Mexico. Application of the system in the tropics is limited. Use of improved genetics (with much British/Continental influence) results in good reproductive rates and increased weaning weights, both of which result in the highest average productivity per cow among all four systems (see table 5). The larger body size and milking ability of a cow means increased nutritional requirements and cattle that are maintained in better body condition year round.

Productivity in this system is relatively high, but management requirements are also high. Increased productivity is accomplished with better grazing management, improved/irrigated pastures, and supplemental feeds. Productivity drops quickly in the nutritionally deficient tropical environments, except in cases of very intensive (and expensive) management. Heat-related challenges in the tropical regions requires the introduction of Zebu genetics, limiting the use of this system for export production in tropical regions.

A "semi-intensive" production system differs from VacaNorte principally because of differences in the final product. Oriented to dual-purpose production or beef production for the domestic market, this system uses more Zebu breeds than the VacaNorte. Calves from this system are generally No. 2 in muscle (less muscle than VacaNorte) and medium to large in frame size. In general, however, this system requires the same level of management as the VacaNorte system, with regard to improved genetics and better health and nutritional management. This system results in slightly larger cattle (compared with traditional production systems) and improved reproductive performance. Overall productivity is less than for VacaNorte systems because of smaller animal size, lighter muscling, and slightly lower reproductive rates. Recently, the increased use of terminal cross beef breeds has created a more heavily muscled calf for domestic markets or for export.

The "traditional" cow-calf system produces a lower quality product than the semi-intensive system, as seen in the slightly lighter muscling and smaller frame size. The principal difference between traditional and semi-intensive production is the level of management. Traditional production is characterized by a wide range of production activities and low levels of health and nutrition management.

Productivity is substantially lower in the traditional system than in semi-intensive, but so too are production costs, with limited use of purchased inputs, especially supplemental feed. The traditional system relies less on genetic improvement and is based on local or regional genetics.

Beef-only and dual-purpose production throughout Mexico may be either semi-intensive or traditional. Depending on the location and resources available, both systems may make considerable use of crop residues and byproduct feeds. In the traditional production system, crop residues (usually from contiguous farming activities) are the primary means by which cattle survive the annual dry season.

Table 5. Characteristics of Mexican cow-calf production systems

	VacaNorte	Semi-intensive	Traditional	Criollo
Cow mature weight	450 kg	435 kg	415 kg	400 kg
Milking ability	Medium/high	Medium	Medium/low	Low
Weaning weight:				
Steers	180 kg	160 kg	140 kg	120 kg
Heifers	162 kg	144 kg	126 kg	108 kg
Average weight	171 kg	152 kg	133 kg	114 kg
Calving (percent)[1]	70-80	60-70	50-60	40-55
Calf death loss (percent)[2]	4	5	8	10
Weaning (percent)[3]	71	61	47	35
Culling (percent)[4]	10	10	8	7
Frame size[5]	Medium/large	Medium/large	Variable	Variable
Muscle thickness[6]	No. 1 – 1 Y2	No. 1 Y2 – 2	No. 2 Y2 – 3	No. 3 – 4
Genetics	High	Medium/high	Medium	Low
Health management[7]	High	Medium/high	Medium/low	Low
Supplementation[8]	Good	Medium	Salt/minerals only	Salt/none
Pasture management	Good	Good/medium	None	None

Source: Derrell Peel, Oklahoma State University.

kg=Kilograms.

[1] The percentage of calves born to cows bred with or exposed to breeding bulls.

[2] The percentage of calves stillborn or that die before weaning.

[3] The percentage of calf crop weaned from the total number of cows bred or exposed, representing a cumulative measure of breeding, calving, and reproductive efficiency in the herd.

[4] The percentage of cows removed from the herd due to death, infertility, low productivity, and advanced age.

[5] Indicates the size of the animal's skeletal structure and has implications for nutrient requirements, reproduction, and other characteristics of animal performance.

[6] The ratio of muscle to bone at a given degree of fatness. Four muscle thickness grades are currently used (No. 1, No. 2, No. 3, and No. 4), from thickest to thinnest, respectively.

[7] Refers to management practices recommended to aid in the prevention and treatment of infectious diseases and in parasite control.

[8] Refers to protein, energy, and/or mineral supplements; the recommended management practice can vary according to production system and forage considerations.

These residues are often very low quality and, without additional (purchased) feed supplements, productivity will be low because the animal's body condition deteriorates sharply when seasonal forage supplies are limited. In semi-intensive production, crop residues are an effective means of lowering the cost of production by serving as a less expensive supplemental feed. Byproduct feeds—including sugarcane bagasse, citrus pulp and waste, brewer's grains, and discarded food products (tortillas, bread, snack chips, etc.)—have been used for many years by cow-calf producers and even more so by the feedlot industry in Mexico. Use of byproduct feeds varies regionally and seasonally, and by type of production system.

"Criollo" production encompasses a wide variety of cattle and activities, but the most critical economic characteristics of this category are low productivity, low management, and low costs of production. With the exception of the rodeo cattle produced in the north, products of this sector are low value or are completely outside the commercial market, with an emphasis on family milk production and draft use. Only occasionally do cull animals provide meat for home consumption. In many cases, Criollo production occurs where crop farming is the primary enterprise. Cattle may be fed a high proportion of crop residues and very little pasture.

Cattle Marketing Systems and U.S. Exports

Cattle marketing practices vary by region and type of production system; however, marketing institutions for live cattle in Mexico are minimal. There is relatively little public reporting of stocker and feeder cattle prices, although there is a national system for reporting slaughter animal and carcass prices. Most stocker and feeder cattle are marketed in some form of direct purchase between buyer and seller, although the primary producer often sells to an intermediary, who is either a speculative trader or a buyer's representative. Most transactions are on cash (or spot) markets with limited use of forward contracts or other marketing arrangements. Price discovery mechanisms are generally informal and producers often sell with little knowledge of prevailing market conditions.

In traditional production systems where cattle are finished on pasture, cattle may be sold only at slaughter or perhaps when the cow-calf producer sells to a local producer to stocker/finish on pasture. In some cases, mostly in the temperate and tropical regions, cattle from larger ranches may be shipped to urban centers for slaughter or to a local buyer who will assemble loads from small producers for shipment to the urban center. Cattle may also be sold directly to a local slaughterhouse, often one or two at a time. As more slaughter facilities are established in more regions, cattle can be sold and slaughtered locally or regionally and moved to major metropolitan markets as chilled carcasses rather than as live animals.

In the principal cattle-exporting regions of the north, cattle for export may be shipped to the border and sold to traders there. Increasingly, large ranches with good quality cattle are selling through U.S. video auctions or directly to stocker or feedlot buyers in the United States. Although not common in most of Mexico, cattle auctions are used in the north, with weekly sales in Chihuahua, Durango, and Hermosillo. Since a high proportion of Mexico's northern cattle that are not exported are fed in feedlots, cattle are increasingly assembled and then shipped long distances from the north by local buyer representatives. In many instances, one or more intermediaries may be involved in such transactions. In some cases, local cattlemen's union facilities may serve as buying stations where producers can bring cattle to be bought by buyer representatives or traders.

Recent dramatic changes in consumer preferences have led to a substantial growth in cattle feeding in confinement and semi-intensive finishing systems using concentrate feeds. An increasing percentage of cattle are transported from southern Mexico to feedlots in northern or central Mexico, increasing shipping distances and costs related to stress on cattle. This trend suggests that there will be increased demand for assembly, transportation, and transfer-of-ownership functions in the marketing system.

OUTLOOK

The Mexican beef industry has been evolving rapidly over the last two decades, primarily as a result of changing consumer dietary preferences and global market impacts. In the 1990s, prolonged drought, coupled with the 1994 peso devaluation and subsequent limitations on capital access, led to Mexican herd liquidation. Cyclically high U.S. cattle prices through much of the next decade resulted in exports of feeder cattle at unsustainable rates. Simultaneously, changing consumer preferences in Mexico resulted in increased demand for beef and changes in the type and quality of meat produced, as well as changes in how it is marketed. Growing beef demand and changing consumer preferences for beef will most likely continue to result in a general increase in beef production in Mexico, especially with respect to cattle in feedlots.

The Mexican beef industry will also face tremendous challenges from both domestic and international competitors. Other domestic livestock industries offer consumers alternative products and compete with beef producers for feed and forage resources. International firms likewise compete against domestic producers for inputs (feed and cattle) and meat. Although beef demand (specifically, demand for fed beef) has increased significantly in recent years, demand for pork and poultry has also increased. The increased demand for feed grains in cattle-feeding operations, along with growth in other livestock sectors and increased competition between food and feed grains, could result in higher grain prices and more marketing outlets.

Such changes have significant implications for cow-calf producers and on overall resource use and production potential in the industry. The increase in total livestock production means greater competition with other types of food and fiber production for limited agricultural resources in the country. This competition will become an increasingly critical factor as the beef industry adopts more intensive production systems to produce fed beef.

Opportunities for improvement in Mexican cattle production, specifically the cow-calf phase, are evident. The adoption of basic animal production and management techniques would greatly improve production efficiencies and the quality of cattle produced, and help to position the Mexican beef industry for the coming decades. The greatest potential for improvement and increased competitiveness in Mexican beef production, however, may lie in improving cattle marketing channels and infrastructure, factors certain to foster efficiency in Mexican cow-calf production techniques.

ACKNOWLEDGMENTS

Thanks go to those who took the time to provide useful comments on early drafts: Don Blayney (Economic Research Service); Jim Robb (former director of Livestock Market Information Center and current vice president of Research and Analysis, CoBank); Eduardo Segarra (professor and chairman, Department of Agricultural and Applied Economics, Texas Tech University); Steven Zahniser (ERS); and Zaida SanJuan (Foreign Agricultural Service, Mexico). Thanks also to our editor, Angela Anderson; our designer, Victor Phillips, Jr.; and to David Marquardt (ERS) whose mapping skills contributed significantly to the final product.

ABOUT THE AUTHORS

Derrell S. Peel is a professor and extension livestock marketing specialist at Oklahoma State University and visiting scholar from Instituto Nacional de Investigaciones Forestales, Agrícolas y Pecuarias (INIFAP) and Fundación Produce, Chihuahua, Mexico. Rachel J. Johnson and Kenneth H. Mathews, Jr., are agricultural economists with USDA's Economic Research Service.

REFERENCES

Cartron, Jean-Luc E., Gerardo Ceballos, and Richard Stephen Felger (eds.). *Biodiversity, Ecosystems, and Conservation in Northern Mexic*o, New York: Oxford University Press, 2005.

FAO (Food and Agriculture Organization). "Country Pasture/Forage Resource Profile: Mexico," *Country Pasture Profiles*, 2005, http://www. fao.org/ag /AGP/agpc/doc/ pasture/forage.htm.

Heath, John Richard. "Evaluating the Impact of Mexico's Land Reform on Agricultural Productivity," *World Development* 20(5): pp. 695-711, 1992.

INEGI (Instituto Nacional de Estadística y Geografía). Censo Agropecuario, 2007, www.inegi.org.mx.

Mejia, Maria, and Derrell S. Peel. "White Corn and Yellow Corn Production in Mexico: Food versus Feed?" *Analysis and Comments* (June 25, 2009, Letter #25), Livestock Marketing Information Center, Denver, Colorado, 2009.

End Notes

[1] Stocker cattle are weaned calves intended for sale as commercial feeder cattle, but have not yet been placed in the feedlot.

[2] Forbs are herbaceous plants that are not grasses.

[3] An animal unit (AU) is 1,000 pounds of grazing animal. A 1,000– pound cow with calf constitutes 1 AU.

[4] Continental and British breeds are two categories of bos taurus breeds of cattle. While both categories originated from Europe, the British breeds, specifically originating from the British Isles, are smaller in size, mature early, and have increased fleshing and marbling ability. The larger Continental breeds are leaner, more muscular, and, as with the British breeds, vary in adaptability to hot climates.

[5] In the minds of many U.S. producers, Zebu is synonymous with Brahman cattle. In fact, Brahman (or more specifically, American Brahman) is only one of several bos indicus breeds used in semi-tropical and tropical climates in the Americas. Zebu breeds used in Mexico include Brahman, Nelore, Gir, Indubrasil, and Guzerat. These breeds share a number of phenotypic characteristics, including the distinctive hump and pendulous hide. However, each of these breeds can be distinguished by color patterns, horn shape, and ear configuration. For example, Indubrasil cattle have extremely long pendulous ears (part of their heat adaptation) and are very distinguishable from Brahman cattle.

[7] In reality, unofficial markets for ejido lands were used almost from the inception of the ejido system. Although not strictly legal, ejido parcels were often leased between ejiditarios and private producers. In the debate over the value of privatizing ejido lands, gains may be limited, in part, because unofficial uses and activities blurred the distinction between ejido and private land (Heath, 1992).

[8] The terms criollo (native) and corriente (common) are often used interchangeably, and each may be used several different ways and have different meanings when applied to various situations. Cattle still exist throughout Mexico that descended from the Spanish cattle originally brought to Mexico in the 16th and 17th centuries. These cattle are properly called criollo or native cattle. In the Sierras of Chihuahua and Durango, for example, these cattle are the corriente or rodeo cattle that are exported to the United States. In the United States, Texas Longhorn cattle and Florida Cracker cattle are remnants of the same Spanish cattle genetics. In other cases, the term criollo may be applied to local cattle from indeterminate genetics, but generally imply cattle of mixed Zebu and dairy genetics. Thus, the definition of criollo cattle depends on the location. These cattle are invariably light muscled with low milking ability and low reproductive rates, but may range from small to large in frame size.

[9] For the last 10 years, the U.S. calf crop as a proportion of the January 1 NASS cattle inventory has averaged 89 percent.

[10] Introduced in 2003, the Programa de Producción Pecuaria Sustentable y Ordenamiento Ganadero y Apícola (PROGAN) seeks to enhance productivity and technology adoption among livestock and apicultural producers. The program provides funding for herd and genetic improvements provided that cattle ranchers meet criteria for animal health standards and management practices.

[11] Spaying heifers is the surgical removal of ovaries (ovariectomy), thus removing the primary source of estrogen and eliminating the ability of the heifer to reproduce. Spaying stocker and feeder heifers is a management tool that maintains them in an "open" state, preventing pregnancy in feedlots and the need to check for pregnancy.

INDEX

G

H

I

Q

R

S